Pro/ENGINEER®
WILDFIRE® 4.0
ESSENTIALS

Kogent Learning Solutions, Inc.

JONES AND BARTLETT PUBLISHERS

Sudbury, Massachusetts

BOSTON   TORONTO   LONDON   SINGAPORE

World Headquarters

Jones and Bartlett Publishers
40 Tall Pine Drive
Sudbury, MA 01776
978-443-5000
info@jbpub.com
www.jbpub.com

Jones and Bartlett Publishers
Canada
6339 Ormindale Way
Mississauga, Ontario L5V 1J2
Canada

Jones and Bartlett Publishers
International
Barb House, Barb Mews
London W6 7PA
United Kingdom

Jones and Bartlett's books and products are available through most bookstores and online booksellers. To contact Jones and Bartlett Publishers directly, call 800-832-0034, fax 978-443-8000, or visit our website, www.jbpub.com.

Substantial discounts on bulk quantities of Jones and Bartlett's publications are available to corporations, professional associations, and other qualified organizations. For details and specific discount information, contact the special sales department at Jones and Bartlett via the above contact information or send an email to specialsales@jbpub.com.

Production Credits

Publisher: David Pallai
Editorial Assistant: Molly Whitman
Production Assistant: Ashlee Hazeltine
Associate Marketing Manager: Lindsay Ruggiero
V.P., Manufacturing and Inventory Control:
    Therese Connell

Composition: diacriTech
Cover and Title Page Design: Scott Moden
Cover Image: © casadaphoto/ShutterStock, Inc.
Printing and Binding: Malloy, Inc.
Cover Printing: Malloy, Inc.

Library of Congress Cataloging-in-Publication Data

Pro/ENGINEER Wildfire 4.0 essentials/Kogent Learning Solutions, Inc.
    p. cm.
    Includes bibliographical references and index.
    ISBN-13: 978-0-7637-8196-5 (pbk.)
    ISBN-10: 0-7637-8196-7 (ibid.)
    1. Pro/ENGINEER. 2. Computer-aided design. 3. Mechanical drawing. I. Kogent Learning Solutions, Inc.
    TA174.P6773 2010
    620'.00420285536–dc22
                                                    2009026046

6048
Printed in the United States of America
13 12 11 10 09   10 9 8 7 6 5 4 3 2 1

TABLE OF CONTENTS

1

INTRODUCING PRO/ENGINEER® WILDFIRE® 4.0

The Pro/ENGINEER Wildfire 4.0 software package is a three-dimensional (3D) computer-aided design (CAD) created by Parametric Technology Corporation (PTC). It is also called a computer-aided manufacturing (CAM) or computer-aided engineering (CAE) software package. It is a feature-based, parametric, and associative solid-modeling software package, which helps a user to create 3D designs for mechanical engineering. It is used to create simple and complex designs faster with improved assembly performance, analyze the design, and automate the transformation of engineering designs into manufacturing processes. You can use it to design a vast range of products, such as soap cases, mobiles, and trains. The product model, during the designing phase, passes through the following modes:

- **Sketch Mode:** Used to sketch geometry for a model that you want to design.
- **Part Mode:** Used to create features of a model.
- **Assembly Mode:** Used to create an assembly design. An assembly design is a design that lets you assemble various parts/features of a model.
- **Drawing Mode:** Used to generate the drawing of the parts and assemblies created in the part and assembly mode.

A model designed in Pro/ENGINEER Wildfire 4.0 incorporates the following modeling approaches:

- **Feature-Based:** A model is made up of a combination of various features such as extrusions, holes, cuts, rounds, sweeps, and slots. A feature is the smallest building block of a model.

- **Parametric:** A model consists of various features that are interrelated, and each feature consists of certain attributes. The parametric approach ensures that modification of a feature's attribute affects the complete model. For example, modifying the diameter of a cylinder in an engine automatically modifies the diameter of the piston.
- **Associative:** A model design in Pro/ENGINEER Wildfire 4.0 passes through various modes, such as **Sketch**, **Part**, and **Assembly**. The associative approach of Pro/ENGINEER Wildfire 4.0 ensures that modifications made in one mode of a model should be reflected in other modes of the same model. For example, if you modify the length of a shaft in the drawing mode, then modifications should also be reflected in the other corresponding modes.
- **Solid Modeling:** The solid modeling approach ensures that the model has properties of a solid object, such as volume and surface area. Solid modeling allows you to create models of physically tangible objects.

1.1 FEATURES OF PRO/ENGINEER WILDFIRE 4.0

In Pro/ENGINEER Wildfire 4.0, many changes and additions have been made to make it more user-friendly than its previous versions. Following are the features of Pro/ENGINEER Wildfire 4.0 that make it popular among CAD/CAM software, such as AutoCAD and GCAM.

- **User Interface:** This feature is divided into many parts, such as message area, navigation pane, browser window, drawing area, top tool chest, and right tool chest. Chapter 2, "Exploring the User Interface," discusses user interface in detail.
- **Auto Round:** This feature automatically rounds the selected edges of a part or an assembly. It saves the time spent in rounding edges manually.
- **Remove Surface:** This feature makes it easy to remove unwanted surfaces. This helps in modifying and simplifying the imported geometries.
- **Shell:** This feature helps you to create shells of various thicknesses for different edges of a part.
- **Dimension Attributes:** This feature appears automatically and allows you to modify dimension attributes of a 3D drawing.
- **View Manager:** This feature helps you to create extract assemblies in Pro/ENGINEER Wildfire 4.0. An extract assembly contains components selected by the user from the parent assembly. The changes made in the

extract assembly are automatically reflected back in the original assembly. An extract assembly is used for concurrent cable designing. Apart from this, the **View Manager** also helps to view a 3D model in different orientations, such as right or left.

- **Surface Manipulation:** This feature makes it easy to modify surfaces simply by picking points on the surface. The surfaces can also be modified by mouse dragging. This feature also allows you to merge surfaces easily by selecting multiple surfaces at a single point of time.

- **Rendering:** This feature lets you view the model in different lighting effects. For example, the **Skylight** feature, which is added newly in Pro/ENGINEER Wildfire 4.0, lets you visualize a model in a skylight environment.

- **Electro-Mechanical Components:** This feature lets you design electro-mechanical components. It also allows you to design ribbon cables.

- **Import Data Doctor:** This feature enables faster import and repair of geometry.

- **Feature Recognition Toolkit:** This feature enables you to import geometries of models. Apart from that, the **Feature Recognition Toolkit** automatically converts the imported geometry into Pro/ENGINEER Wildfire 4.0 features such as holes, rounds, and chamfer.

1.2 FREQUENTLY USED TERMS IN PRO/ENGINEER WILDFIRE 4.0

Some of the commonly used terms of Pro/ENGINEER Wildfire 4.0 are as follows:

- **Entity:** An entity is defined as the basic element of a feature. An entity can be a line, an arc, a spline, or a circle.
- **Dimension:** A dimension in Pro/ENGINEER Wildfire 4.0 is defined as the measurement of entities.

Note: The measurement unit in Pro/ENGINEER Wildfire 4.0 is mm.

- **Parameter:** A parameter is defined as a numeric value that defines a feature, such as thickness. A parameter can be modified any time.
- **Relation:** A relation in Pro/ENGINEER Wildfire 4.0 can be defined as an equation used to relate two entities.

- **Constraints:** In Pro/ENGINEER Wildfire 4.0, constraints are defined as logical operations that are performed on selected entities to make them accurate with respect to other entities. For example, the equal constraint in Pro/ENGINEER Wildfire 4.0 is used to make selected entities equal in size.
- **Weak Dimensions and Weak Constraints:** In Pro/ENGINEER Wildfire 4.0, weak dimensions and weak constraints appear automatically on entities in a sketch. They are considered as temporary dimensions and constraints. They appear in gray color.
- **Strong Dimensions and Strong Constraints:** The dimensions and constraints that are added manually to a sketch are called strong dimensions and strong constraints. They appear in white color and are not deleted automatically.

1.3 INSTALLING PRO/ENGINEER WILDFIRE 4.0

To design models with Pro/ENGINEER Wildfire 4.0 software, it must be installed on your computer. Table 1.1 lists the basic system requirements for the installation of Pro/ENGINEER Wildfire 4.0:

Type of Requirement	Value
Operating System	Windows Vista Business Edition, Windows Vista Ultimate Edition, Windows XP Professional Edition, Windows XP Home Edition, or UNIX
Hard Disk	Minimum: 2.0 GB Recommended: 3.0 GB or higher
CPU Speed	Minimum: 500 MHZ Recommended: 2.4 GHz or higher
RAM	Minimum: 256 MB Recommended: 1024 MB or higher
SWAP Space	Minimum: 500 MB Recommended: 2048 MB or higher
Browser Support	Microsoft Internet Explorer 6.0 or higher
Monitor Resolution	1024 x 768 screen resolution with 24-bit color support; however, higher resolution with higher color support is recommended
Mouse	Microsoft-approved three-button mouse
Network	Microsoft-approved Ethernet Network Adapter or Network Card

TABLE 1.1   System requirements to install Pro/ENGINEER Wildfire 4.0

You can install the Pro/ENGINEER Wildfire 4.0 software on a license-server machine, a license-client machine, or on both. However, to install Pro/ENGINEER Wildfire 4.0 software either on a license-server machine or on a license-client machine, you have to first install FLEXnet Publisher. FLEXnet Publisher is a software-license manager, which is a product of Acresso Software. FLEXnet Publisher is used in corporate environments to provide floating licenses to users attached with a license-server machine. If you install the Pro/ENGINEER Wildfire 4.0 software only on the license-client machine, the license-client machine works as both server and client.

Note: A floating license allows multiple users to share an application installed on a server machine.

In this section, we see the process to install the Pro/ENGINEER Wildfire 4.0 software on a license-client machine on Windows Vista Ultimate Edition. Let's now follow these steps to install Pro/ENGINEER Wildfire 4.0:

1. *Insert* the DVD containing the Pro/ENGINEER Wildfire 4.0 software into the DVD ROM drive of your computer. The **AutoPlay** window appears, as shown in **Figure 1.1**.

Note: If the **AutoPlay Window** does not appear, then access it from your computer's DVD drive.

---

2. Now, *click* the **Open** folder to view the files folder icon in the **General options** section to explore the Pro/ENGINEER Wildfire 4.0 software component, as shown in Figure 1.1:

FIGURE 1.1

The Pro/ENGINEER Wildfire 4.0 software component appears.

3. *Copy* the **license.dat** file anywhere in your computer. The file is required to be copied because later on, if you update the Pro/ENGINEER Wildfire 4.0 software with the latest version, then you will also have to update the **license.dat** file.

4. Now, *double-click* the **PTC.Setup** installation utility to start the process of installing the Pro/ENGINEER Wildfire 4.0 software. The PTC welcome screen appears, as shown in **Figure 1.2**:

FIGURE 1.2

The welcome screen disappears within a few seconds and the Pro/ENGINEER Wildfire 4.0 setup window appears, as shown in **Figure 1.3**.

5. *Click* the **Next** button, as shown in Figure 1.3:

FIGURE 1.3

The license agreement window appears (**Figure 1.4**).

6. *Select* the check box beside the "I Accept the Agreement Terms and Conditions" label to accept the software agreement terms and conditions.

Note: If you do not agree with the software terms and conditions, then select the check box beside "I Decline the Agreement Terms and Conditions" to quit installation.

7.   *Click* the **Next** button, as shown in Figure 1.4:

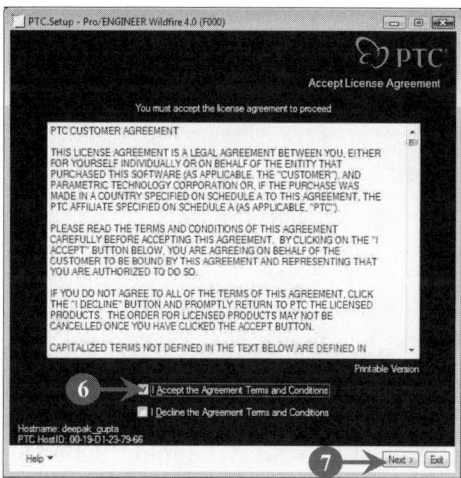

FIGURE 1.4

The **Select Product to be installed** window appears (**Figure 1.5**).

8.   *Click* the **Pro/ENGINEER & Pro/ENGINEER Mechanica** link, as shown in Figure 1.5:

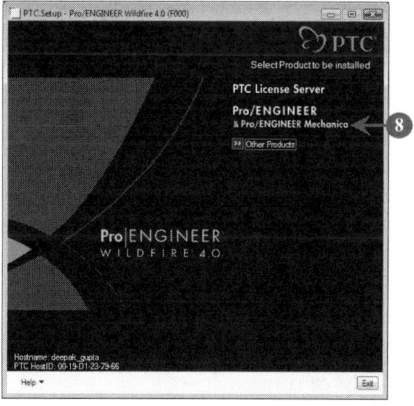

FIGURE 1.5

The **Define Installation Components** window appears, as shown in **Figure 1.6**:

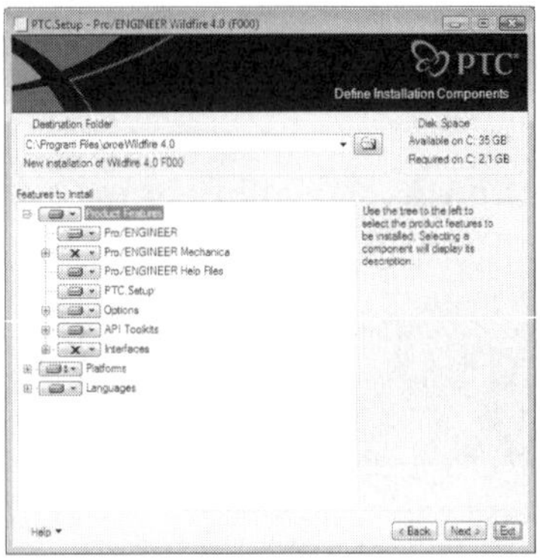

FIGURE 1.6

On this window, you see the following options:

- **Destination Folder:** It is the folder where the Pro/ENGINEER Wildfire 4.0 software is installed. By default, the destination folder is created in the **Program Files** folder. However, you can change the location of the destination folder by *clicking* the folder icon shown in Figure 1.6. If you change the destination folder, then you must take care that the folder in which the software is to be installed must be empty or newly created.
- **Disk Space:** This option informs you about the total space available on the hard disk and the required space to install the software.
- **Features to Install:** This option lets you select the features and sub-features of the software that you want to install. It contains the following main components:
- **Product Features:** Contains the following features that can be installed:
  ◊ **Pro/ENGINEER:** Installs the files required to run the Pro/ENGINEER software.
  ◊ **Pro/ENGINEER Mechanica:** Helps an engineer to understand the product performance in a better way. It also helps in producing good quality products at low cost.

Note: By default, this feature is disabled. The process to enable this feature is discussed ahead in this section.

  ◊ **Pro/ENGINEER Help Files:** Installs the Pro/ENGINEER help files in the destination folder.
  ◊ **PTC Setup:** Used to reconfigure the software after installation without executing the **PTC.Setup** utility.

Note: To reconfigure the software configuration after installation, go to the **Destination folder > bin** and execute the **ptcsetup** file. The **ptcsetup** file is a batch file that needs administrator privileges to execute.

◊ **Options:** Installs **ModelCHECK** configurator, which is used to validate the completeness of 3D drawings, and **Mold Component Catalog**, which is used for mold tooling.

◊ **API Toolkits:** Installs the files required to run the Application Program Interface Toolkit, such as Visual Basic API and Pro/TOOLKIT.

◊ **Interfaces:** Installs interface tools for third-party software, such as **Pro/INTERFACE for CATIA** (Computer-Aided Three-Dimensional Interactive Application) and **Pro/INTERFACE for JT**.

● **Platforms:** Selects your system's default platform to install Pro/ENGINEER Wildfire 4.0 software.

● **Languages:** Lets you select and install the language for the software. By default, the English language is installed. This feature cannot be disabled.

The **Red Cross** mark beside the features **Pro/ENGINEER Mechanica** and **Interfaces** (Figure 1.6) implies that they are disabled to be installed. To install these features, you first need to enable them.

9. To enable the **Pro/ENGINEER Mechanica** feature, *right-click* the feature. A flyout appears, as shown in **Figure 1.7**.

10. *Select* the **Install all sub-features** option on the flyout (Figure 1.7):

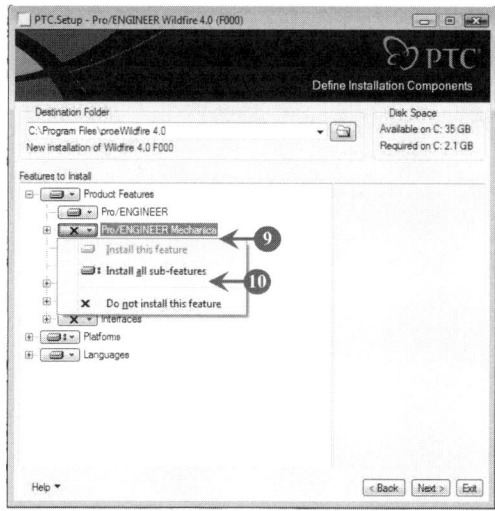

FIGURE 1.7

11. *Repeat* steps 9 and 10 to enable the **Interfaces** feature. Both the features along with their subfeatures are now enabled (**Figure 1.8**).

12. *Click* the **Next** button to continue the process, as shown in Figure 1.8:

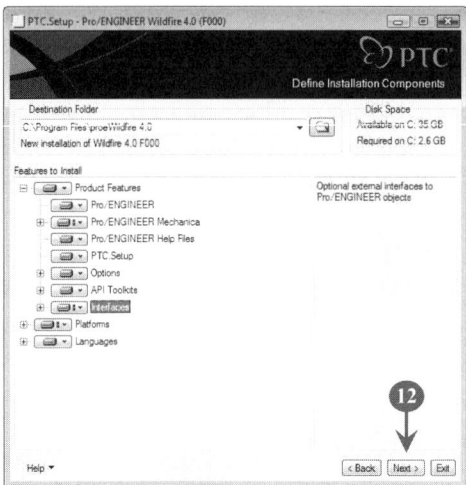

FIGURE 1.8

The **FLEXnet license servers** window appears, as shown in **Figure 1.9**:

13. Now, to add the **license.dat** file, *click* the **Add** button in the **FLEXnet license servers** window (Figure 1.9):

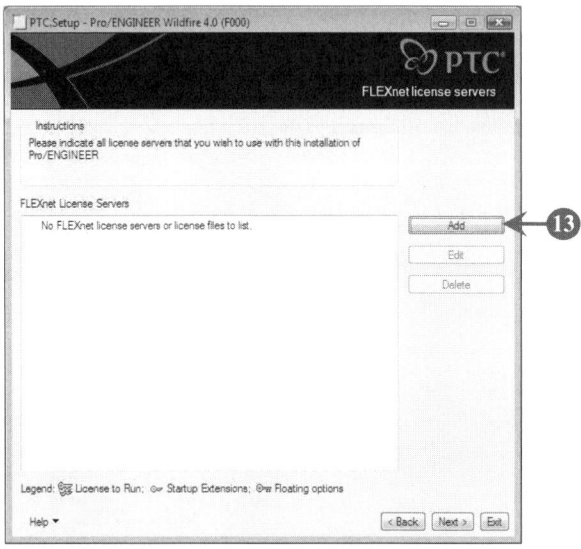

FIGURE 1.9

The **Specify License Server** dialog box appears, as shown in **Figure 1.10**.

14. *Select* the radio button beside the **Locked license file (no server running)** label (Figure 1.10).

15. *Click* the folder icon, as shown in Figure 1.10:

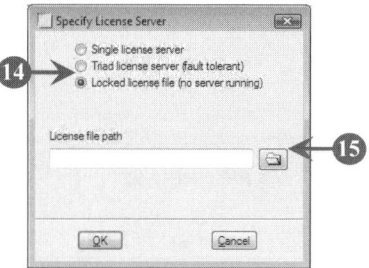

FIGURE 1.10

The **Select File** dialog box appears, as shown in **Figure 1.11**.

16. *Select* the **license.dat** file (Figure 1.11).

17. *Click* the **Open** button, as shown in Figure 1.11:

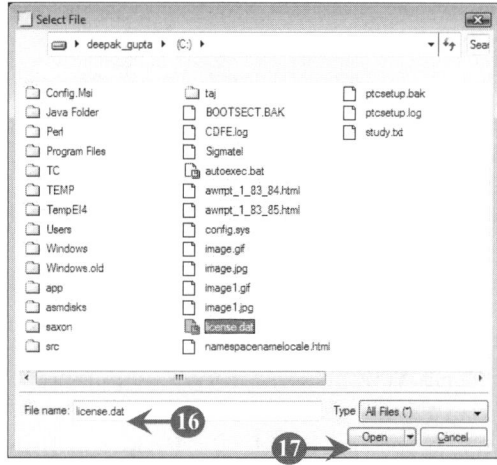

FIGURE 1.11

The path of the license file appears in the **Specify License Server** dialog box, as shown in **Figure 1.12**.

18. *Click* the **OK** button, as shown in Figure 1.12:

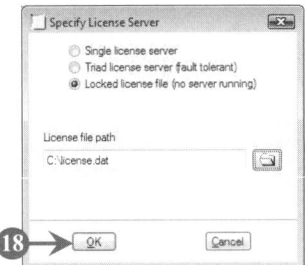

FIGURE 1.12

The **license.dat** file is added in the **FLEXnet server** window, as shown in **Figure 1.13**.

19. *Click* the **Next** button, as shown in Figure 1.13:

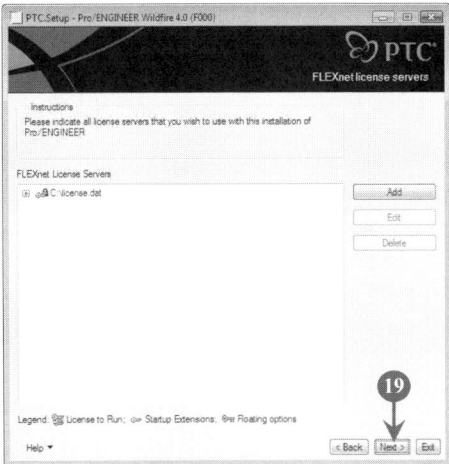

FIGURE 1.13

The Pro/ENGINEER Wildfire 4.0 window, which helps you create a shortcut, appears as shown in **Figure 1.14**.

By default, the check box beside the **Program folder** label is selected. If you want to create shortcut icons in the start menu and on the desktop, select the check boxes beside the **Desktop** and **Start menu** labels. In our case, we are continuing with the default settings.

20. *Click* the **Next** button, as shown in Figure 1.14:

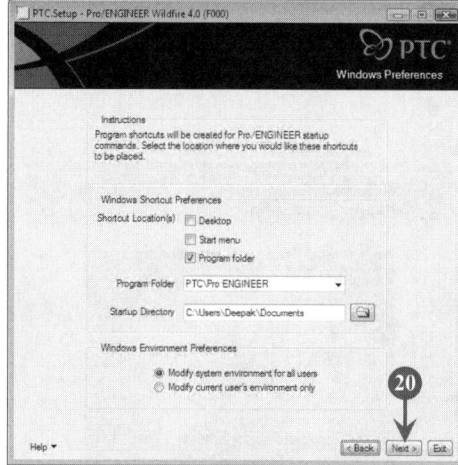

FIGURE 1.14

The Pro/ENGINEER Wildfire 4.0 window to choose additional configuration options appears, as shown in **Figure 1.15**.

21. *Click* the **Install** button (Figure 1.15) to continue the installation process:

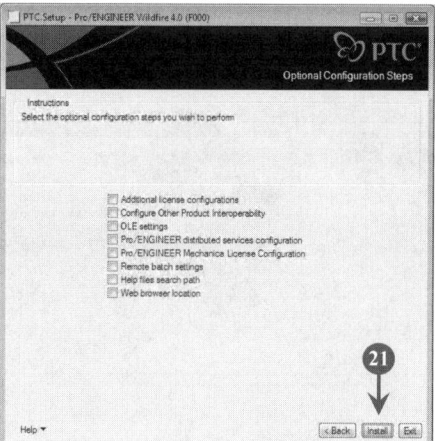

FIGURE 1.15

In order to configure any additional option, select the check box beside that option. In our case, we are not selecting additional options.

The **Installation Progress** window appears, as shown in **Figure 1.16**.

Buttons on Figure 1.16 are disabled at this moment. These buttons get enabled when the installation progress completes.

22. *Click* the **Next** button, as shown in Figure 1.16:

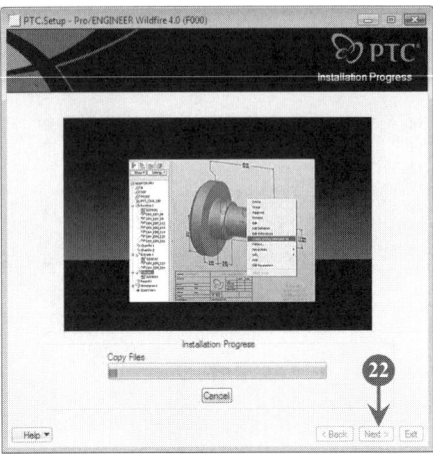

FIGURE 1.16

The Pro/ENGINEER Wildfire 4.0 software is now installed on your system. The Pro/ENGINEER Wildfire 4.0 exit window appears, as shown in **Figure 1.17**.

23. *Click* the **Exit** button (Figure 1.17) to exit from the installation wizard:

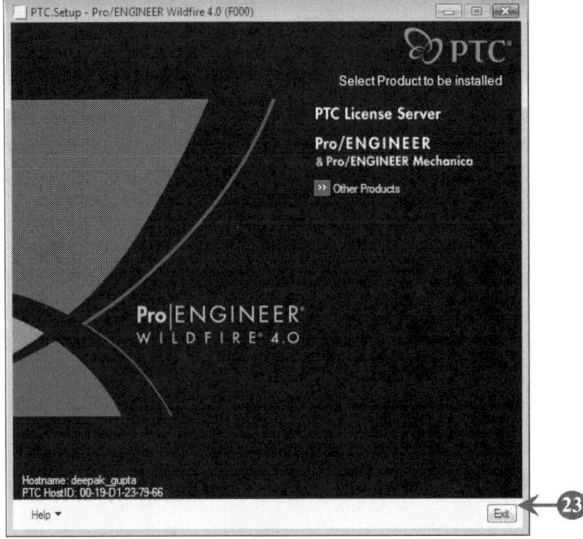

FIGURE 1.17

With this, the process to install Pro/ENGINEER Wildfire 4.0 is completed.

24. *Click* the **Start** > **All Programs** > **PTC** > **Pro ENGINEER** > **Pro ENGINEER** option. The Pro/ENGINEER Wildfire 4.0 initial screen appears, as shown in **Figure 1.18**:

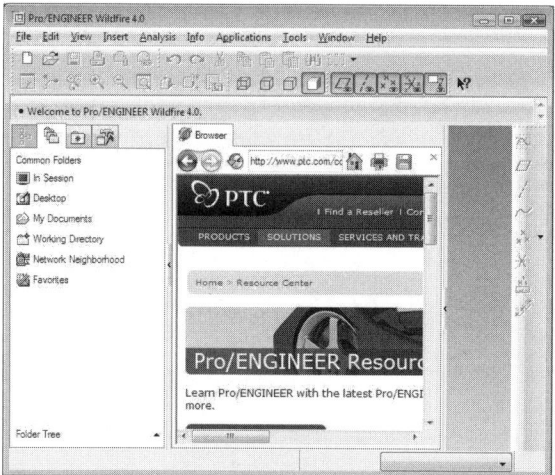

FIGURE 1.18

Now, you can start working with Pro/ENGINEER Wildfire 4.0. Let's now summarize the main points covered in this chapter.

SUMMARY

In this chapter, you have learned about:

- Features of Pro/ENGINEER Wildfire 4.0.
- Frequently used terms in Pro/ENGINEER Wildfire 4.0.
- Installing Pro/ENGINEER Wildfire 4.0 on a client machine.

2

EXPLORING THE USER INTERFACE

In This Section
◇ Starting Pro/ENGINEER Wildfire 4.0
◇ Performing File Management

Interface is a medium that allows interaction between a program and a user. To understand an interface, consider a scenario where you need to operate a television set. To operate a television set, you need a remote control to perform various functions such as changing the channels, controlling the volume, changing the display settings, and so on. In this case, the remote control acts as an interface between you and the television set. Similarly, in computing, an interface is required for the interaction between the user and the software that is commonly known as user interface (UI). The UI of Pro/ENGINEER Wildfire 4.0 also acts as a medium between users and the software to help users in designing real-world products such as cars, cell phones, and automobile parts. In order to work with software, it is important to understand the fundamentals of its UI. Therefore, let's start by discussing the UI of Pro/ENGINEER Wildfire 4.0 in this chapter.

This chapter explains the use of each menu on the menu bar of Pro/ENGINEER Wildfire 4.0 and proceeds with a discussion about various toolbars. The chapter also discusses managing files in Pro/ENGINEER Wildfire 4.0. Let's now see how to start Pro/ENGINEER Wildfire 4.0.

2.1 STARTING PRO/ENGINEER WILDFIRE 4.0

Once you have installed Pro/ENGINEER Wildfire 4.0 successfully on your system, open it by *clicking* **Start** > **All Programs** > **PTC** > **Pro ENGINEER** > **Pro ENGINEER**, as shown in **Figure 2.1**:

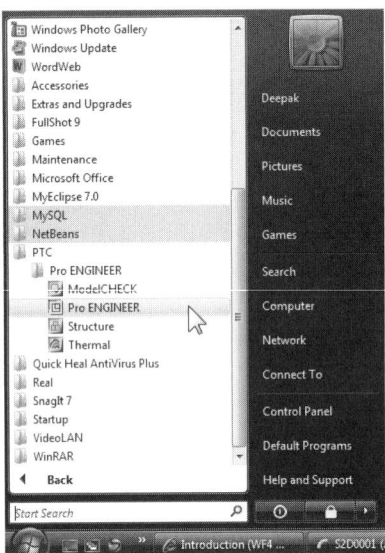

FIGURE 2.1

Note: The process of installing the Pro/ENGINEER Wildfire 4.0 software is discussed in Chapter 1, "Introducing Pro/ENGINEER Wildfire 4.0."

Pro/ENGINEER Wildfire 4.0 on start-up displays a screen, as shown in **Figure 2.2**:

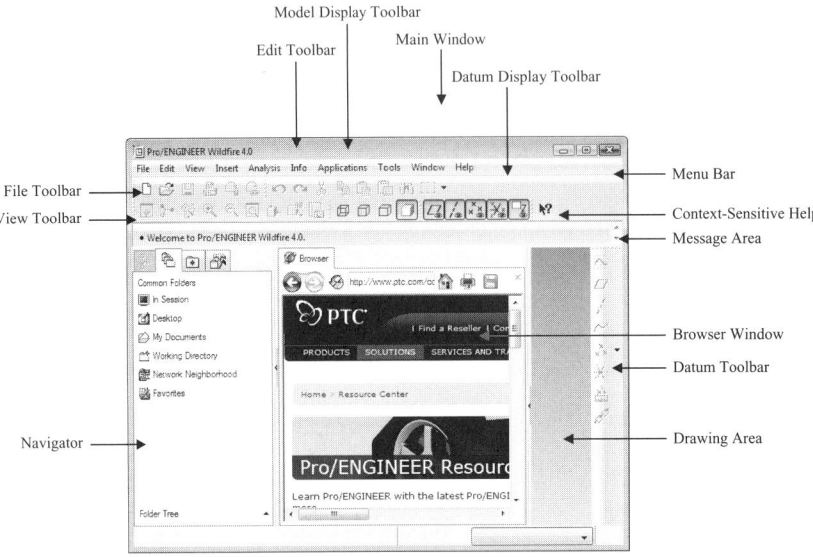

FIGURE 2.2

Figure 2.2 shows various windows, menu bar, and toolbars in the start-up screen of Pro/ENGINEER Wildfire 4.0. When you start Pro/ENGINEER Wildfire 4.0, the window that appears is known as the **Main Window**. All other windows are located within the **Main Window**.

Following are the components that are located within the **Main Window** of Pro/ENGINEER Wildfire 4.0:

- Message Area
- Browser Window
- Drawing Area
- Navigator
- Menu Bar
- Toolbars

Message Area

Message Area displays messages regarding the current status of the operation performed. The **Message Area** also displays information to complete an operation. The **Message Area** is shown in **Figure 2.3**:

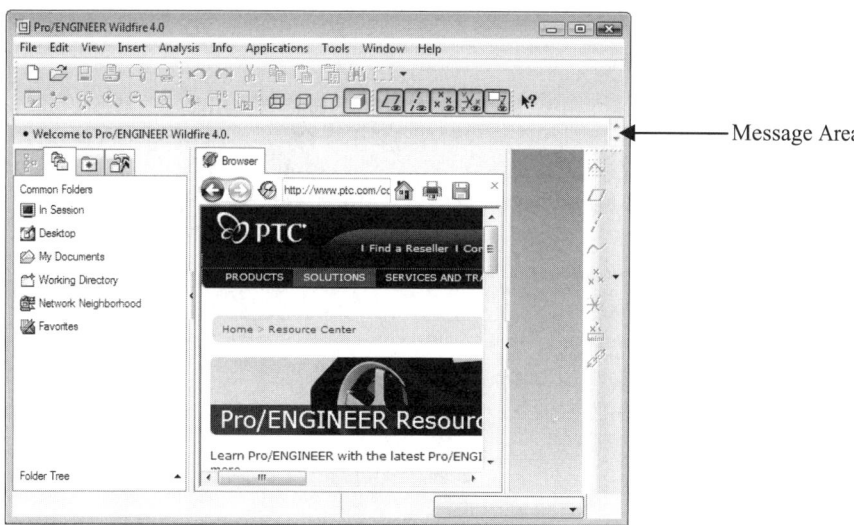

FIGURE 2.3

Figure 2.3 shows the **Message Area** displaying a message "Welcome to Pro/ENGINEER Wildfire 4.0." to the user.

Browser Window

The **Browser Window** is used to preview the model of any existing Pro/ENGINEER Wildfire 4.0 file.

When Pro/ENGINEER Wildfire 4.0 is started, the **Browser Window** is displayed on the screen. However, we can hide the **Browser Window** by *clicking* the **sash** control (⌷). The **Browser Window** is shown in **Figure 2.4**:

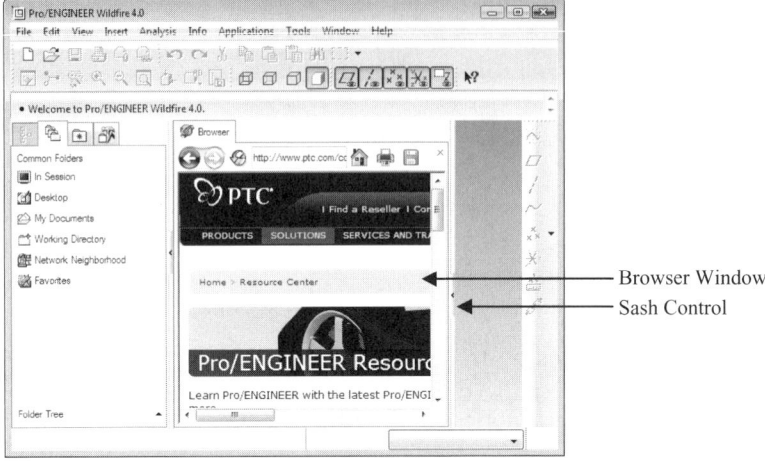

FIGURE 2.4

Drawing Area

Drawing Area is the screen where a Pro/ENGINEER Wildfire 4.0 model is created. The sketching of various entities takes place in the **Drawing Area**. The size of the **Drawing Area** can be adjusted using the sash controls located on the edges of the **Navigator** and the **Browser Window**. The **Drawing Area** is shown in **Figure 2.5**:

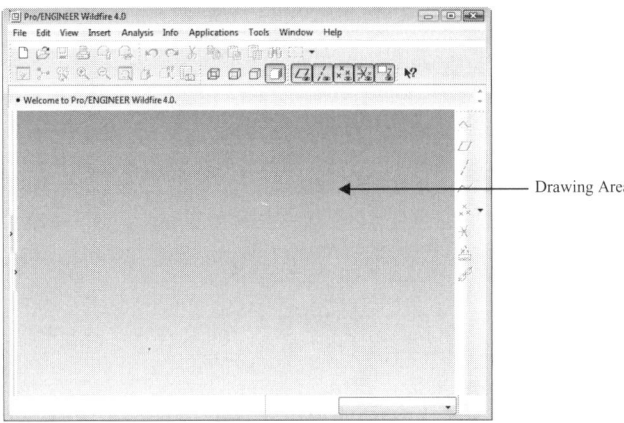

FIGURE 2.5

Navigator

Navigator is located to the left of the **Drawing Area** and contains multiple tabs for performing functions, such as browsing the Pro/ENGINEER Wildfire 4.0 files, displaying information about a specific model.

Figure 2.6 shows the multiple tabs available in the **Navigator**:

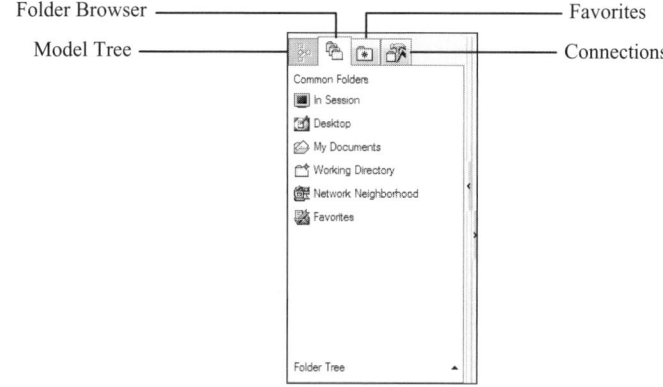

FIGURE 2.6

The tab options shown in Figure 2.6 are described in the following section:

- **Model Tree:** The **Model Tree** tab ▦ is used for manipulating the features in Pro/ENGINEER Wildfire 4.0 Part files. The manipulation includes deleting, suppressing, renaming, and editing the features such as cut, hole, and shell. However, the **Model Tree** also allows the user to manipulate assembly components in Assembly files. Manipulating features in Part files is discussed in Chapter 4, "Exploring Pro/ENGINEER Wildfire 4.0 Part Mode."
- **Folder Browser:** The **Folder Browser** tab ▦ allows the user to browse and open folders located in the local system. In addition to these functions, the **Folder Browser** also helps in creating and setting a folder as the working directory. *Click* the **Folder Browser** tab located on the **Navigator** and a list of folders is displayed, as shown in **Figure 2.7**:

FIGURE 2.7

- **Favorites:** The **Favorites** tab  allows user to save the frequently used directory folders under the **Personal Favorites** folder. The **Personal Favorites** folder is located under the **Favorites** tab, as shown in **Figure 2.8**:

FIGURE 2.8

- **Connections:** The **Connections** tab  is used to access the Parametric Technology Corporation (PTC) services and other websites within the environment of Pro/ENGINEER Wildfire 4.0. *Click* the **Connections** tab to select a list of options, such as Browser, 3DModelSpace, User Area, User Group, PTC.com, and Customer Support, as shown in the **Figure 2.9**:

FIGURE 2.9

Menu Bar

The **Menu Bar**, shown in Figure 2.2, contains many important menu options. Following are the menus of Pro/ENGINEER Wildfire 4.0:

- File menu
- Edit menu
- View menu
- Insert menu
- Analysis menu
- Info menu
- Applications menu
- Tools menu
- Window menu
- Help menu

The File Menu

The **File menu** contains several options used for manipulating the files. To access options available in the **File menu**, *select* **File** from the **Menu Bar** and a list of options is displayed, as shown in **Figure 2.10**:

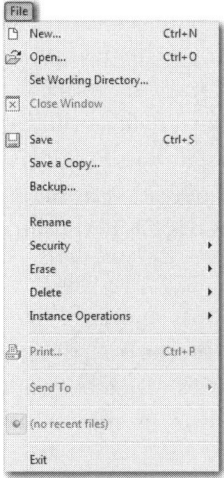

FIGURE 2.10

Some of the commonly used **File menu** options are as follows:

- **New:** Creates a new file, such as Part file, Assembly file, and Drawing file.
- **Open:** Opens an existing Pro/ENGINEER Wildfire 4.0 file saved in the local system.
- **Set Working Directory:** Selects any directory of the local system for placing the files created in Pro/ENGINEER Wildfire 4.0 during a session. In case a working directory is not selected, the files are saved in the default directory.

Note: The default directory is selected during the installation of the Pro/ENGINEER Wildfire 4.0 software.

- **Save:** Saves a file that exists in the current session of Pro/ENGINEER Wildfire 4.0.

Note: In Pro/ENGINEER Wildfire 4.0, multiple files can exist in a particular session. A file continues to exist in a session until it is erased from the memory.

- **Save a Copy:** Saves a copy of a file in any location of the local system or in the current working directory. This option is mainly used for saving a file in other file formats, such as STEPS and STL.
- **Backup:** Creates a backup of a file existing in the current session of Pro/ENGINEER Wildfire 4.0.
- **Rename:** Changes the name of a file that exists in the current session.
- **Erase:** Erases files from the temporary memory of Pro/ENGINEER Wildfire 4.0. The temporary memory contains files that are not yet saved permanently onto the hard disk of the local system.
- **Delete:** Deletes files created in Pro/ENGINEER Wildfire 4.0 permanently from the hard disk of the local system.
- **Exit:** Exits the Pro/ENGINEER Wildfire 4.0 environment.

The Edit Menu

The options available under the **Edit menu** are used for modifying Pro/ENGINEER Wildfire 4.0 elements, such as entities. To access options available in the **Edit menu**, *select* **Edit menu** from the **Menu Bar**, as shown in **Figure 2.11**:

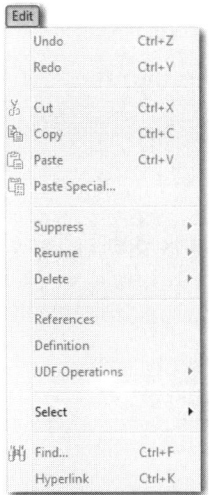

FIGURE 2.11

Some of the commonly used **Edit menu** options are as follows:

- **Undo:** Returns the state of a file from the current state to the previous state.
- **Redo:** Reverses the action made by the **Undo** option in Pro/ENGINEER Wildfire 4.0 file.

- **Cut:** Removes the selected entity, such as line, arc, or circle, from the **Drawing Area**.
- **Copy:** Duplicates the selected entity from the **Drawing Area** to place it on the desired location.
- **Paste:** Inserts the entity copied or removed from the **Drawing Area** using the **Copy** or **Cut** option.
- **Paste Special:** Pastes the original feature or a set of features on the desired location of the **Drawing Area**.

The View Menu

The options available under the **View menu** are used to modify the appearance of a model and work screen of Pro/ENGINEER Wildfire 4.0. To access options available in the **View menu**, *select* **View menu** from the **Menu Bar**, as shown in **Figure 2.12**:

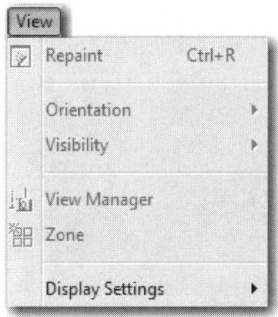

FIGURE 2.12

Some of the commonly used **View menu** options are as follows:

- **Repaint:** Removes temporary information from the **Drawing Area**.
- **Orientation:** Orients a model created in a Pro/ENGINEER Wildfire 4.0 session. Orientation of a model includes zooming or enlarging, spinning, or panning the selected model. Panning of a model refers to the horizontal or vertical movement of a model,
- **View Manager:** Creates extract assemblies in Pro/ENGINEER Wildfire 4.0. An extract assembly contains components selected by the user from the parent assembly. The changes made in the extract assembly are automatically reflected back in the original assembly. An extract assembly is used for concurrent cable designing.
- **Display Settings:** Changes the appearance and the color scheme of the model and the **Drawing Area**. It contains the following options:

- **Model Display:** Controls the appearance of the model on the **Drawing Area**.
- **Datum Display:** Displays datums created in a Pro/ENGINEER Wildfire 4.0 session. Datums are features that work as a reference for creating a feature. Datums are discussed in detail in Chapter 4, "Exploring Pro/ENGINEER Wildfire 4.0 Part Mode."
- **System Colors:** Displays the default system colors. These colors can be allocated to various elements of a Pro/ENGINEER Wildfire 4.0 model.
- **Performance:** Improves the display quality during dynamic orientation of a shaded model.

The Insert Menu

The **Insert menu** provides direct access to the commands for creating Pro/ENGINEER Wildfire 4.0 features, such as datum planes, holes, and protrusion. However, these commands are enabled while working in the Part mode. Features are discussed in Chapter 4, "Exploring Pro/ENGINEER Wildfire 4.0 Part Mode". To access options available in the **Insert menu**, select **Insert menu** from the **Menu Bar**, as shown in **Figure 2.13**:

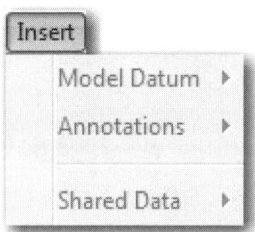

FIGURE 2.13

Some of the commonly used **Insert menu** options are as follows:

- **Model Datum:** Creates various datum features, such as datum plane, datum axis, datum curve, datum points, and system coordinates.
- **Annotations:** Creates annotation elements in a model. Annotation elements in Pro/ENGINEER are data features that are used to share model information among various models.

The Analysis Menu

The options available under the **Analysis menu** are used to perform functions, such as locating properties of Assembly and Part files. To access options available in the **Analysis menu**, select **Analysis menu** from the **Menu Bar**, as shown in **Figure 2.14**:

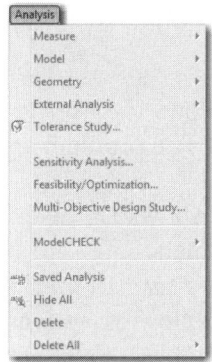

FIGURE 2.14

Some of the commonly used **Analysis menu** options are as follows:

- **Tolerance Study:** Performs tolerance analysis in a current working session. Tolerance analysis is used to predict the effects of manufacturing defects using Pro/ENGINEER Wildfire 4.0.
- **Sensitivity Analysis:** Performs sensitivity analysis that allows a user to identify the effect of changing the range of certain dimensions in a model. Usually standardized fixed values are specified to the dimensions.
- **Feasibility/Optimization:** Provides optimized dimension values to overcome certain user-defined constraints in a model.
- **Multi-Objective Design Study:** Provides a multi-objective design study to help users to explore several options for designing the model in Pro/ENGINEER Wildfire 4.0.
- **Saved Analysis:** Saves an analysis along with the model. The **Saved Analysis** option is also used to create an analysis feature from a saved analysis. An analysis feature is used to automate the design process in Pro/ENGINEER Wildfire 4.0.

The Info Menu

The **Info menu** provides information regarding the creation of a model that is available in the session. The information includes parent–child relationships, features, and geometry of the model. To access options available in the **Info menu**, select **Info menu** from the **Menu Bar**, as shown in **Figure 2.15**:

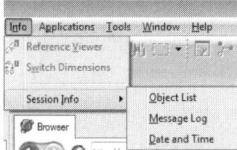

FIGURE 2.15

Some of the commonly used **Info menu** options are as follows:

- **Reference Viewer:** Allows the user to graphically view a complex model to locate problem areas. The problem areas generally appear after a feature is modified in the model.
- **Switch Dimensions:** Allows the user to switch between numeric- and symbolic-dimension formats.
- **Session Info:** Allows the user to manage a Pro/ENGINEER Wildfire 4.0 session. It contains the following options:
  - **Object List:** Displays a list of all the models that are opened in a current working session. This option also allows the user to perform manipulations such as saving, editing, and closing the list containing the models.
  - **Message Log:** Displays a list of messages received in a current session. These messages include error messages resulting from regeneration failures. A regeneration failure is an error that occurs when a specified model has failed to generate in a session. The **Message Log** option also allows the user to perform manipulations such as saving, editing, and closing the list containing the messages.
  - **Date and Time:** Displays the current date and time in the **Message Area** of Pro/ENGINEER Wildfire 4.0.

The Applications Menu

The **Applications menu** contains options to access manuals provided by PTC. The user can also switch between various modes of the design process available in Pro/ENGINEER Wildfire 4.0 using the **Applications menu**.

Note: The options under the **Applications menu** are not discussed here, as these options are not available at this stage. Options under the **Applications menu** are displayed once we start working in any of the Pro/ENGINEER Wildfire 4.0 modes.

The Tools Menu

The **Tools menu** contains options for customizing the UI of the Pro/ENGINEER Wildfire 4.0. The customization of the UI includes changing the appearance of the **Drawing Area**, creating mapkeys, and so on. To access options available in the **Tools menu**, select **Tools menu** from the **Menu Bar**, as shown in **Figure 2.16**:

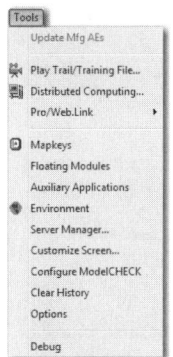

FIGURE 2.16

Some of the commonly used **Tools menu** options are as follows:

- **Play Trail/Training File:** Allows the user to run a trail file. A trail file is a collection of all selected menu options and keyboard entries of a specific Pro/ENGINEER Wildfire 4.0 working session.
- **Distributed Computing:** Allows the user to enable a hardware to get accessed by multiple computers over a network.
- **Mapkeys:** Allows the user to create, delete, view, and save mapkeys in the configuration file, **mapkey**. A mapkey is a collection of frequently used commands that are executed in sequence using a specified key on the keyboard.
- **Floating Modules:** Allows the user to obtain a floating module license. A floating module license allows Pro/ENGINEER Wildfire 4.0 to run on any system on a network.
- **Auxiliary Applications:** Allows the user to manage auxiliary applications in Pro/ENGINEER Wildfire 4.0. Auxiliary applications are those applications that are created outside the environment of Pro/ENGINEER Wildfire 4.0. The managing of auxiliary applications includes registering, starting, stopping, and deleting the auxiliary applications.
- **Environment:** Allows the user to change the environment settings for the current working session of Pro/ENGINEER Wildfire 4.0.

The Window Menu

The **Window menu** allows the user to open, close, and resize the Pro/ENGINEER Wildfire 4.0 window. Pro/ENGINEER Wildfire 4.0 allows you to open multiple windows of a single model or different models simultaneously; however, the user is allowed to work only in a single window at a time. The user can switch among multiple windows by selecting the desired window from the list of open windows under the **Window menu**.

To access options available in the **Window menu**, select **Window menu** from the **Menu Bar**, as shown in **Figure 2.17**:

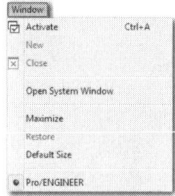

FIGURE 2.17

Some of the commonly used **Window menu** options are as follows:

- **Activate:** Used to make the selected window active. The window in which the user is currently working is known as the active window. The active window enables the file in it to utilize all the features supported by Pro/ENGINEER Wildfire 4.0.
- **New:** Creates a new window containing the same file that is present in the parent window. The parent window is the one from which the new window is created.
- **Close:** Closes the current window. However, if any object is present in that window, it is retained in the memory.
- **Maximize:** Used to increase the size of the current window to its full size.
- **Default Size:** Used to resize the current window to its default size.

The Help Menu

The **Help menu** allows the user to access Help information on using Pro/ENGINEER Wildfire 4.0. It provides context-sensitive help options to locate information about Pro/ENGINEER Wildfire 4.0 menus. To access options available in the **Help menu**, select **Help menu** from the **Menu Bar**, as shown in **Figure 2.18**:

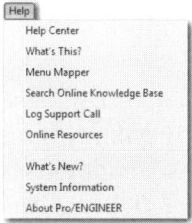

FIGURE 2.18

Some of the commonly used **Help menu** options are as follows:

- **Help Center:** Allows the user to access help on any Pro/ENGINEER Wildfire 4.0 topic. The Pro/ENGINEER Wildfire 4.0 **Help** window opens after selecting this option.
- **What's This?:** Enables context-sensitive help. The context-sensitive help displays information about the selected menu options.
- **Menu Mapper:** Opens a tool named **Menu Mapper**, which helps the user to find the menu options available in all the previous versions of Pro/ENGINEER, such as Pro/ENGINEER Wildfire 3.0, Pro/ENGINEER Wildfire 2.0, and Pro/ENGINEER Wildfire 2001.
- **Online Resources:** Allows the user to access online information on various Pro/ENGINEER Wildfire 4.0 topics.
- **What's New?:** Allows the user to access new functionalities introduced with Pro/ENGINEER Wildfire 4.0, such as the latest tools, menus, and tutorials.

Toolbars

The toolbars contain buttons that are shortcut commands to the options available under the **Menu Bar**. These toolbars help users access the menu options with a click of a button on the toolbar.

In Figure 2.2, the toolbars are located at two locations on the **Main window**. The first location where toolbars, such as the **File toolbar**, **View toolbar**, **Model Display toolbar**, and **Datum Display toolbar** are located is called the **Top tool chest**. The **Top tool chest** is present at the top of the **Drawing Area** in the **Main window**. The second location is at the right side of the **Drawing Area**; this location is known as the **Right tool chest**. The toolbar located at this position is known as **Datum toolbar**.

Pro/ENGINEER Wildfire 4.0 provides the following toolbars:

- File Toolbar
- Edit Toolbar
- View Toolbar
- Model Display Toolbar
- Datum Display Toolbar
- Datum Toolbar

The File Toolbar

The **File toolbar** contains buttons that are shortcut commands to some of the commonly used **File menu** options, as shown in **Figure 2.19**:

FIGURE 2.19

The buttons available in the **File toolbar** are as follows:

- **New:** The **New** button ▫ allows the user to create a new Pro/ENGINEER Wildfire 4.0 file.
- **Open:** The **Open** button ☞ allows the user to open an existing Pro/ENGINEER Wildfire 4.0 file.
- **Save:** The **Save** button ▣ allows the user to save a Pro/ENGINEER Wildfire 4.0 file created in the current working session.
- **Print:** The **Print** button ▤ allows the user to print a selected model created in the current session.
- **Mail Recipient (as Link):** The **Mail Recipient (as Link)** button ◪ allows the user to send a link of the current model with an e-mail.
- **Mail Recipient (as Attachment):** The Mail Recipient (as Attachment) button ◪ allows the user to send the current model as an attachment with an e-mail.

The Edit Toolbar

The **Edit toolbar** contains buttons that are shortcut commands to some of the commonly used **Edit menu** options. The buttons available under the **Edit toolbar** are used to modify the entities created in a Pro/ENGINEER Wildfire 4.0 session. **Figure 2.20** shows the **Edit toolbar**:

FIGURE 2.20

The buttons available in the **Edit toolbar** are as follows:

- **Undo:** The **Undo** button ↺ returns the state of a file from current to previous state.
- **Redo:** The **Redo** button ↻ reverses the action made by the **Undo** command in the Pro/ENGINEER Wildfire 4.0 file.
- **Cut:** The **Cut** button ✄ removes the selected entity, such as a line, arc, or circle from the **Drawing Area**.
- **Copy:** The **Copy** button ▤ duplicates the selected entity from the **Drawing Area** to place it on the desired location.
- **Paste:** The **Paste** button ▤ inserts the entity copied or removed from the **Drawing Area** using the **Copy** or **Cut** button.
- **Paste Special:** The **Paste Special** button ▤ pastes the original feature or a set of features on the desired location of the **Drawing Area**.

- **Find: The Find** button ▦ allows the user to search for entities from the model created in a session.
- **Inside Box:** The **Inside Box** button ▦ selects drawing items, such as points and triangles, that lie completely inside the rectangular region. *Click* the ▪ button on the **Edit toolbar** to explore the remaining buttons. **Figure 2.21** shows the remaining buttons available on the **Edit toolbar:**

FIGURE 2.21

Following are the remaining buttons available under the **Edit toolbar:**

- **Across Box:** The **Across Box** button ▦ selects drawing items that lie either partially or completely inside the rectangular region.
- **Inside Polygon:** The **Inside Polygon** button ⬚ selects drawing items that lie completely inside the polygon region.
- **Paintbrush:** The **Paintbrush** button ✎ selects drawing items on which the stroke of a paintbrush can be applied.
- **Inside Ellipse:** The **Inside Ellipse** button ⬭ selects drawing items inside the ellipse region.

The View Toolbar

The **View toolbar** contains buttons that are shortcut commands to some of the commonly used **View** Menu options, such as **Repaint** and **View Manager**. **Figure 2.22** shows the **View Toolbar:**

FIGURE 2.22

The buttons available in the **View Toolbar** are as follows:

- **Repaint:** The **Repaint** button ▨ allows the user to repaint the **Drawing Area**. Repainting is meant for removing temporary information from the **Drawing Area**.

- **Spin Center:** The **Spin Center** button ⛝ allows the user to enable or disable the **Spin Center**. The **Spin Center** is located at the center of rotation in a model.

- **Orient Mode:** The **Orient Mode** button ▨ allows the user to enable or disable the visibility of the datums. The **Orient Mode** also allows the user to provide more capability than the standard spinning, panning, and zooming techniques.

- **Zoom In:** The **Zoom In** button ⚲ allows the user to enlarge the selected entity on the **Drawing Area**. You can manually enlarge an entity by using the scroll key of the mouse.

- **Zoom Out:** The **Zoom Out** button ⚲ allows the user to reduce the view of the selected entity to its half on each click of this button.

- **Refit:** The **Refit** button ▨ allows the user to adjust the size of an entire sketch such that it can be seen completely on the **Drawing Area**.

- **Reorient:** The **Reorient** button ▨ allows the user to orient a model created in a Pro/ENGINEER Wildfire 4.0 session. The orientation of the view includes setting the orientation center besides spinning, panning, and zooming the model.

- **Named View List:** The **Named View List** button ▨ allows the user to display saved views. The saved views shows the changed orientation when the default orientation of the model changes.

- **View Manager:** The **View Manager** button ▨ allows the user to create extract assemblies in Pro/ENGINEER Wildfire 4.0. An extract assembly contains components selected by the user from the parent assembly. The changes made in the extract assembly are automatically reflected back in the original assembly. Extract assembly is used for concurrent cable designing.

The Model Display Toolbar

The **Model Display toolbar** contains buttons that are shortcut commands to the options present under the submenu of the **View menu** option called **Display Settings**. These buttons are used for controlling the appearance of a model on the **Drawing Area**. **Figure 2.23** shows the **Model Display toolbar**:

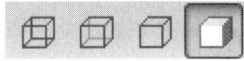

FIGURE 2.23

The buttons available in the **Model Display toolbar** are as follows:

- **Wireframe:** The **Wireframe** button 🔲 is used to show hidden lines, such as edges on the selected model. These hidden lines are shown in white color.
- **Hidden line:** The **Hidden line** button 🔲 is used to display hidden lines in gray color.
- **No hidden:** The **No hidden** button 🔲 is used to remove hidden lines from the current model.
- **Shading:** The **Shading** button 🔲 is used to display the selected model as shaded. The **Shading** button is applied by default to a model in a session.

The Datum Display Toolbar

The **Datum Display toolbar** contains shortcut commands to the options under the submenu of the **View menu** option called **Display Settings**. These shortcut commands are executed using buttons available in the **Datum Display toolbar**. These buttons are mainly used to control the visibility of some datum features, such as datum planes, datum axes, and datum points. **Figure 2.24** shows the **Datum Display toolbar**:

FIGURE 2.24

The buttons available in the **Datum Display toolbar** are as follows:

- **Plane Display:** The **Plane Display** button 🔲 allows the user to display datum planes along with their names.
- **Axis Display:** The **Axis Display** button 🔲 allows the user to display datum axes along with their names.
- **Point Display:** The **Point Display** button 🔲 allows the user to display datum points along with their names.
- **Csys Display:** The **Csys Display** button 🔲 allows the user to display coordinate systems along with their names. Coordinate systems are used in Pro/ENGINEER Wildfire 4.0 to perform tasks, such as detailed element analysis and modeling applications.
- **Annotation Element Display:** The **Annotation Element Display** button 🔲 allows the user to display annotation elements along with their names. Annotation elements in Pro/ENGINEER Wildfire 4.0 are data features that are used to share model information among various models.

The Datum Toolbar

The **Datum toolbar** contains buttons for creating various datum features, such as datum plane, datum axis, and datum curve. The datum features created using the **Datum toolbar** act as base features to create other features in Pro/ENGINEER Wildfire 4.0. **Figure 2.25** shows the buttons available on the **Datum toolbar**:

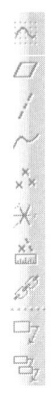

FIGURE 2.25

Figure 2.25 shows buttons available in the **Datum Display toolbar**. Let's discuss these buttons one by one:

- **Sketch:** The **Sketch** button is used to sketch datum curves. The sketched datum curves consist of one or more sketch segments.
- **Datum Plane:** The **Datum Plane** button is used to create datum planes, in addition to the three datum planes available by default in every Pro/ENGINEER Wildfire 4.0 session.
- **Datum Axis:** The **Datum Axis** button is used to create datum axes in a model. The datum axes are like the centerline needed at the center of orthographic models, such as holes and cylinders.
- **Datum Curve:** The **Datum Curve** button is used to create datum curves in a model. The datum curves are useful in creating advanced solid features in Pro/ENGINEER Wildfire 4.0.
- **Datum Point:** The **Datum Point** button is used to create datum points in a model. The datum points are useful in creating Pipe features.
- **Coordinate System:** The **Coordinate System** button is used to create coordinate systems in a model.
- **Analysis Feature:** The **Analysis Feature** button is used to create analysis features in a model.

- **Reference:** The **Reference** button ✐ is used to create datum reference features in a Pro/ENGINEER Wildfire 4.0 session. Datum reference features are user-defined datum planes, datum axes, and datum points.
- **Annotation Feature:** The **Annotation Feature** button ▣ is used to create annotation features in a model.
- **Annotation Element Propagate:** The **Annotation Element Propagate** button ▣ is used to insert an annotation element propagate feature in a model.

2.2 PERFORMING FILE MANAGEMENT

The objects that are created in a Pro/ENGINEER Wildfire 4.0 session are stored in the form of files. The user can perform manipulations, such as creating, opening, saving, and deleting these files. Pro/ENGINEER Wildfire 4.0 provides ways for manipulating files similar to those in the Windows operating system. However, there are major differences between managing files in Pro/ENGINEER Wildfire 4.0 and the Windows operating system. The following is a list of differences in managing files by Pro/ENGINEER Wildfire 4.0 and the Windows operating system:

- The rules for naming files in Pro/ENGINEER Wildfire 4.0 are more stringent than those for Windows. Pro/ENGINEER Wildfire 4.0 filenames are limited up to 31 characters and cannot contain periods, brackets, and spaces. However, files created in the Windows operating system do not comply with any complex file-naming conventions.
- Each time a Pro/ENGINEER Wildfire 4.0 file is saved, a new version of that file is created. However, in the Windows operating system, the original version of the file gets updated each time the file is saved.

Before exploring the various options available to manipulate Pro/ENGINEER Wildfire 4.0 files, let's discuss some of the general file extensions available in Pro/ENGINEER Wildfire 4.0.

Understanding the General File Extensions

As discussed earlier, the objects created in a Pro/ENGINEER Wildfire 4.0 session are stored in the form of files. These files are saved with certain file extensions. Pro/ENGINEER Wildfire 4.0 supports various file extensions based on the working mode being selected by the user. **Table 2.1** lists the file extensions in Pro/ENGINEER Wildfire 4.0:

Mode	Extension
Sketch	*.sec
Part	*.prt
Assembly	*.asm
Drawing	*.drw

TABLE 2.1   General file extensions in Pro/ENGINEER Wildfire 4.0

Note: Various Pro/ENGINEER Wildfire 4.0 modes are discussed in Chapter 1, "Introducing Pro/Engineer Wildfire 4.0."

Introducing the Working Directory

Pro/ENGINEER Wildfire 4.0 allows the user to select any directory on the local system as a working directory to save all the files created during a session. A working directory should be selected before starting work in Pro/ENGINEER Wildfire 4.0; else, the files will be saved in the default directory. The default directory is selected during the installation of the Pro/ENGINEER Wildfire 4.0.

Following are the two ways to select the working directory in Pro/ENGINEER Wildfire 4.0:

- Using the Navigator.
- Using the Select Working Directory Dialog Box.

Using the Navigator

The **Navigator** is a pane that is displayed on the left side of the **Drawing Area** and contains multiple tabs for performing functions, such as browsing the Pro/ENGINEER Wildfire 4.0 files and displaying information about a specific model. The Navigator can be used to select a folder as the working directory for a Pro/ENGINEER Wildfire 4.0 session. The detailed discussion on various tabs available under the **Navigator** has already been given in the section, "Navigator", of this chapter.

Perform the following steps to select a working directory using the **Navigator**:

1. *Click* the **Folder Tree** button on the **Navigator**, as shown in **Figure 2.26**:

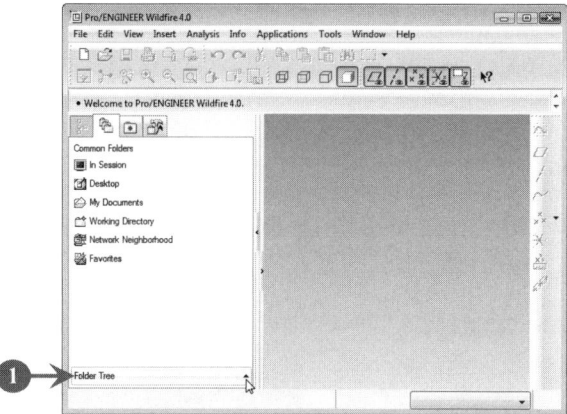

FIGURE 2.26

A list of folders and directories available under the local system gets displayed (**Figure 2.27**).

2. *Right-click* any folder, say **ProE**, on the local system, as shown in Figure 2.27.

The shortcut menu appears displaying some **File menu** options.

3. Now, *select* the **Set Working Directory** option from the shortcut menu to set the selected folder as the working directory (Figure 2.27):

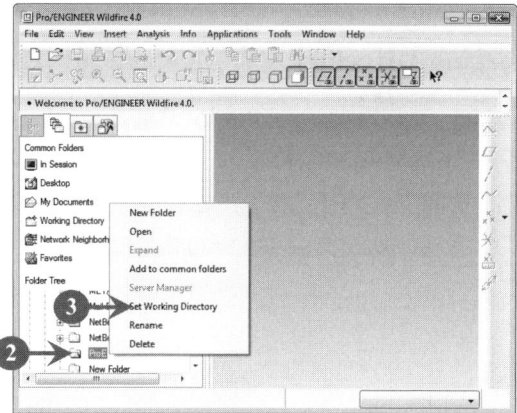

FIGURE 2.27

The **ProE** folder is set as a working directory for a Pro/ENGINEER Wildfire 4.0 session.

Using the Select Working Directory Dialog Box

Another way to set a folder as the working directory is by using the **Select Working Directory** Dialog Box.

Perform the following steps to select a working directory using the **Select Working Directory** Dialog Box:

1. *Select* the **File>Set Working Directory** option from the **Menu Bar**, as shown in **Figure 2.28**:

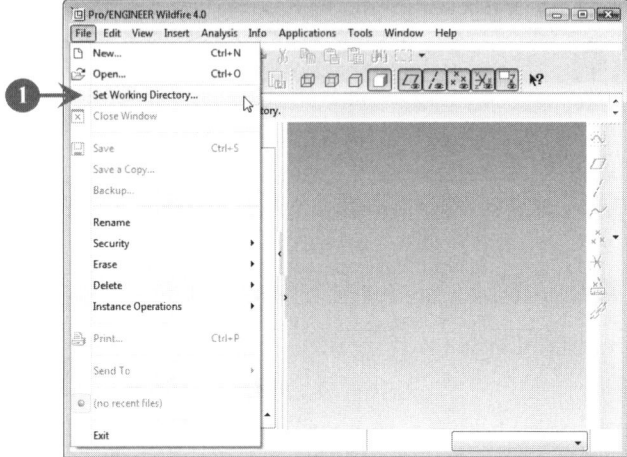

FIGURE 2.28

The **Select Working Directory** dialog box appears (**Figure 2.29**).

2. *Select* any folder, say **ProE**, on the local system and *click* the **OK** button to set it as the working directory, as shown in Figure 2.29:

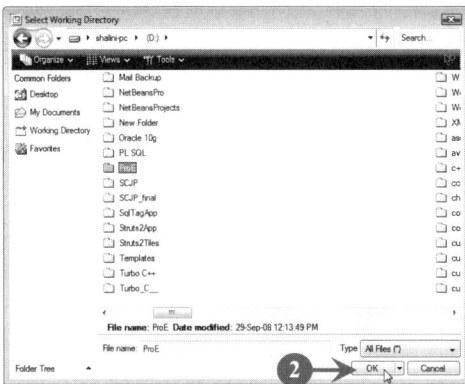

FIGURE 2.29

Once the working directory for a Pro/ENGINEER Wildfire 4.0 session is specified, you can start working with Pro/ENGINEER Wildfire 4.0. Let's start with creating a file in Pro/ENGINEER Wildfire 4.0.

Creating a File

The user can create files of any type from any mode of Pro/ENGINEER Wildfire 4.0. The file created in Pro/ENGINEER Wildfire 4.0 can be a **Sketch** file, **Part** file, **Assembly** file, and **Drawing** file.

Perform the following steps to create a file in Pro/ENGINEER Wildfire 4.0:

1. *Start* Pro/ENGINEER Wildfire 4.0. The process to start Pro/ENGINEER Wildfire 4.0 has already been discussed in the section, "Starting Pro/ENGINEER Wildfire 4.0" of this chapter.

2. *Select* the **File > New** option from the **Menu Bar**, as shown in **Figure 2.30**:

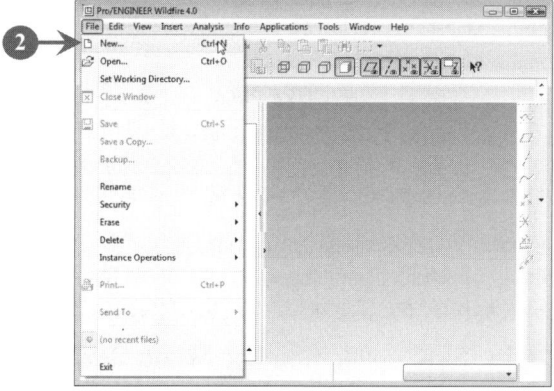

FIGURE 2.30

The **New** dialog box appears displaying the various modes available in Pro/ENGINEER Wildfire 4.0. The **Part** mode radio button is selected by default. A default name of the **Part** file is also shown in the **Name** text box (**Figure 2.31**).

Note: The User default template text box is selected in the **New** dialog box. This creates a new file using a default template including three default datum planes.

3.   *Click* the **OK** button in the **New** dialog box, as shown in **Figure 2.31**:

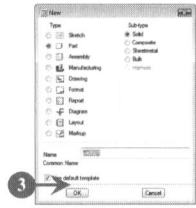

FIGURE 2.31

An initial screen containing the three default datum planes as a result of entering the **Part** mode is shown in **Figure 2.32**:

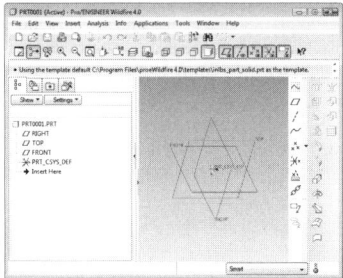

FIGURE 2.32

Saving a File

The **Save** option allows user to save files of any type from any mode of Pro/ENGINEER Wildfire 4.0.

Perform the following steps to save a file in Pro/ENGINEER Wildfire 4.0:

1.   *Select* the **File** > **Save** option from the **Menu Bar**, as shown in **Figure 2.33**:

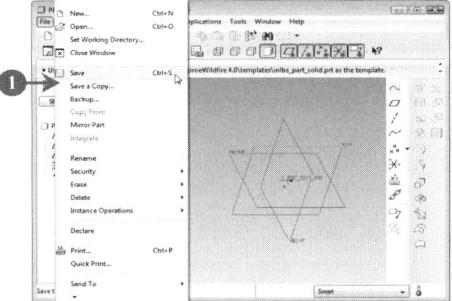

FIGURE 2.33

The **Save Object** dialog box appears and the default file name of the object is shown in the **Model Name** text box (**Figure 2.34**).

2. *Click* the **OK** button from the **Save Object** dialog box to save the file, as shown in Figure 2.34:

FIGURE 2.34

You can also save a copy of the current file in any location of the local system or in the current working directory using the **Save a Copy** option.

Perform the following steps to save a copy of the current file in Pro/ENGINEER Wildfire 4.0:

1. *Select* the **File > Save a Copy** option from the **Menu Bar**, as shown in **Figure 2.35**:

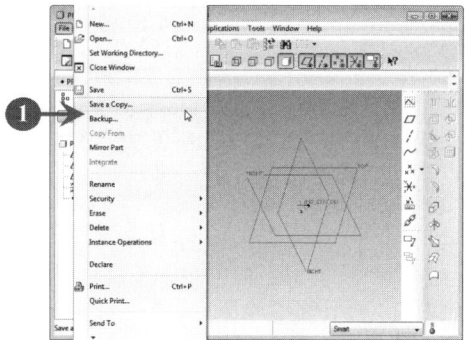

FIGURE 2.35

The **Save a Copy** dialog box appears, and the file name of the object is shown in the **Model Name** box along with the target directory to place a copy of the file. The **New Name** text box is provided to specify a new name for the file to be saved.

2. *Click* the **OK** button from the **Save a Copy** dialog box to save the file, as shown in **Figure 2.36**:

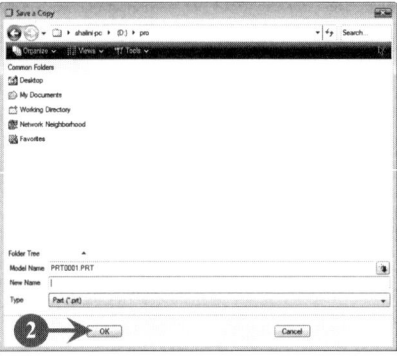

FIGURE 2.36

Figure 2.36 shows that a copy of the file, PRT0001.PRT, is created outside the working directory. The copy gets created in the folder named **pro**.

Backing up a File

The **Backup** option allows the user to create a backup of a file that exists in a session.

Perform the following steps to create a backup of a file in Pro/ENGINEER Wildfire 4.0:

1. *Select* the **File > Backup** option from the **Menu Bar**, as shown in **Figure 2.37**:

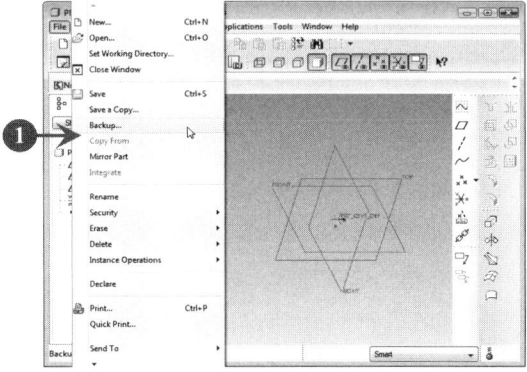

FIGURE 2.37

The **Backup** dialog box appears (**Figure 2.38**), and the file name of the object for which the backup is to be taken is shown in the **Model Name** box.

2. *Select* the target directory (backup) to place the backup of the file in the **Backup** dialog box. The name of the selected directory is placed automatically in the **Backup To Text** box (Figure 2.38).

3. *Click* the **OK** button from the **Backup** dialog box to create the backup of the file, as shown in Figure 2.38:

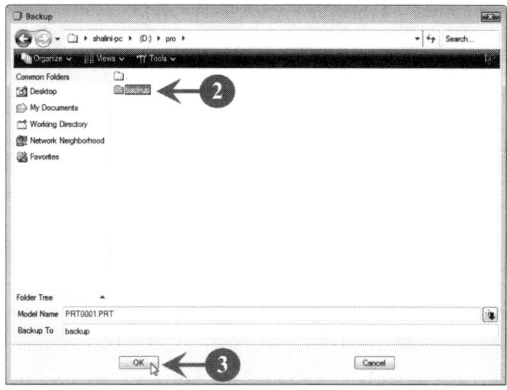

FIGURE 2.38

A backup of the file PRT0001.PRT is created in a folder named backup.

Renaming a File

The **Rename** option allows the user to change the name of a Pro/ENGINEER Wildfire 4.0 file.

Perform the following steps to change the name of an existing file in Pro/ENGINEER Wildfire 4.0:

1. *Select* the **File > Rename** option from the **Menu Bar**, as shown in **Figure 2.39**:

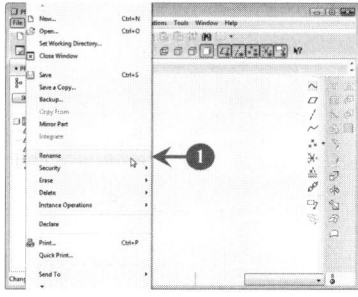

FIGURE 2.39

The **Rename** dialog box appears to provide a new name for a file (**Figure 2.40**).

2. *Specify* a new name to the file (PRT0001.PRT) in the **New Name** text box (Figure 2.40). In our case, we have specified **DemoPart** as a new name.

3. *Click* the **OK** button from the **Rename** dialog box to rename the file, as shown in Figure 2.40:

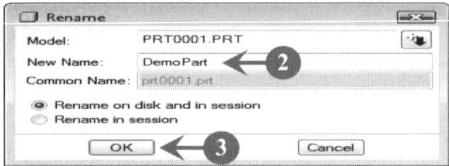

FIGURE 2.40

Note: The **Rename on disk and in session** radio button is selected by default in the **Rename** dialog box to rename the file on the disk as well as in the current session.

Deleting a File

The user can permanently delete files created in Pro/ENGINEER Wildfire 4.0 from the hard disk of the local system by using the **Delete** option from the **File menu**.

Perform the following steps to delete a file in Pro/ENGINEER Wildfire 4.0:

1. *Select* the **File > Delete > Old Versions** option from the **Menu Bar**, as shown in **Figure 2.41**:

FIGURE 2.41

A **Message Input Window** is displayed in the **Message Area** to accept the name of the file whose old versions are to be deleted (**Figure 2.42**). In the

Message Input Window, specify the name of a file whose version you want to delete. In our case, the file name is PRT0001.PRT.

2. *Click* the **Accept Value** (✓) button on the **Message Input Window** to delete the old versions of the selected file, as shown in Figure 2.42:

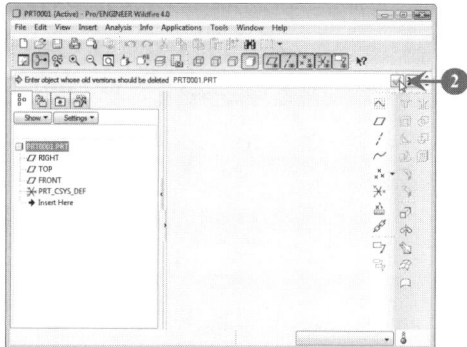

FIGURE 2.42

Note: The **File** > **Delete** > **All Versions** option from the **Menu Bar** deletes all the versions of the selected file from the local system.

Opening an Existing File

The user can open an existing Pro/ENGINEER Wildfire 4.0 file by browsing the local system using the **Open** option from the **File menu** or **Open** button from the **File toolbar**.

Perform the following steps to open a file in Pro/ENGINEER Wildfire 4.0:

1. *Select* the **File** > **Open** option from the **Menu Bar** as shown in **Figure 2.43**:

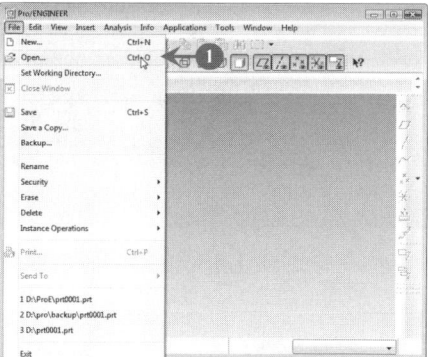

FIGURE 2.43

The **File>Open** dialog box appears and displays a list of Pro/ENGINEER Wildfire 4.0 files.

2.   *Select* the appropriate file and *click* the **Open** button from the **File>Open** dialog box to open the file, as shown in **Figure 2.44**:

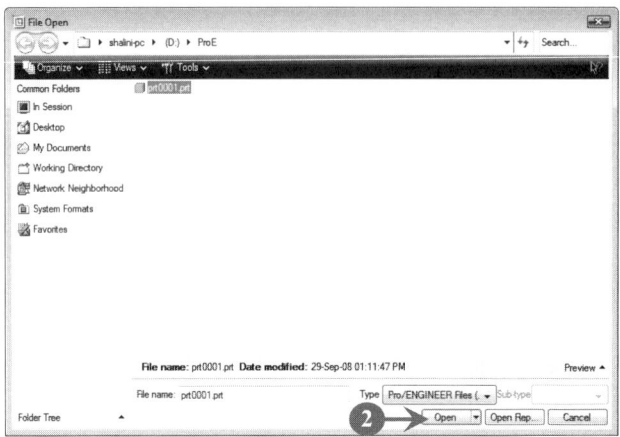

FIGURE 2.44

3.   The selected file (prt0001.prt) opens, as shown in **Figure 2.45**:

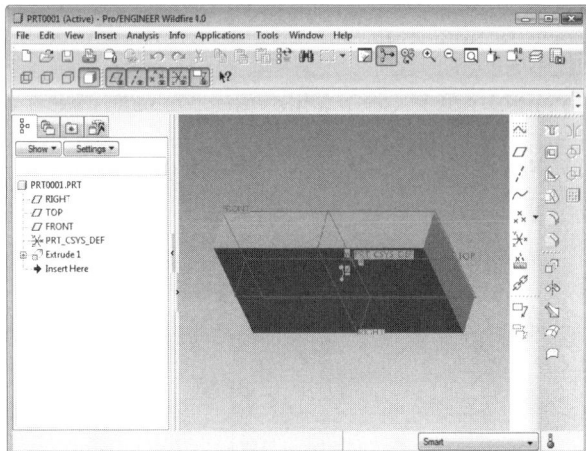

FIGURE 2.45

SUMMARY

In this chapter, you have learned about:

- Starting Pro/ENGINEER Wildfire 4.0.
- Various components available within the Main Window of the Pro/ENGINEER Wildfire 4.0.
- Menu options available under the **Menu Bar**.
- Various toolbars available in Pro/ENGINEER Wildfire 4.0.
- Performing file management in Pro/ENGINEER Wildfire 4.0.

Chapter 3

EXPLORING PRO/ENGINEER WILDFIRE 4.0 SKETCH MODE

In This Section

◊ Starting the Sketch Mode
◊ Sketcher Tools Toolbar
◊ Working with Sketching Entities
◊ Working with Dimensions
◊ Applying Constraints
◊ Modifying Dimensions
◊ Trimming Entities
◊ Mirroring Entities
◊ Scaling and Rotating Entities
◊ Importing 2D Drawings
◊ Deleting Entities

Sketch mode in Pro/ENGINEER Wildfire 4.0 is used to draw a two-dimensional (2D) sketch of a three-dimensional (3D) feature. For example, to create a rectangular slot, you first draw its 2D sketch and then that sketch is converted into a 3D rectangular slot in **Part** mode. To draw a sketch in Pro/ENGINEER Wildfire 4.0, various sketching entities such as line, circle, and rectangle are used. These sketching entities are drawn by using various buttons, such as line button and rectangle button, both available on the Sketcher Tools toolbar in Sketch mode. A combination of various sketching entities leads to a complete sketch. However, to create a sketch of a feature, we first have to invoke the **Sketch** mode. Therefore, we start this chapter by invoking or starting the **Sketch** mode. Then, we depict the **Sketcher Tools toolbar** that helps you draw various sketching entities. Once a sketch is drawn, that sketch needs to be dimensioned and constrained. A sketch will be complete only when it is fully dimensioned and constrained. You may also need to modify dimensions of an entity or to delete an entity. We discuss the process of modifying dimensions and deleting an entity at the end of this chapter.

3.1 STARTING THE SKETCH MODE

To start the **Sketch** mode, you first have to start the Pro/ENGINEER Wildfire 4.0. The process to start the Pro/ENGINEER Wildfire 4.0 is discussed in Chapter 2, *"Exploring the User Interface"*. Let's now follow the steps given below to start the **Sketch** mode:

1. *Select* the **File menu** on the menu bar in the Pro/ENGINEER Wildfire 4.0 main window, as shown in **Figure 3.1**:

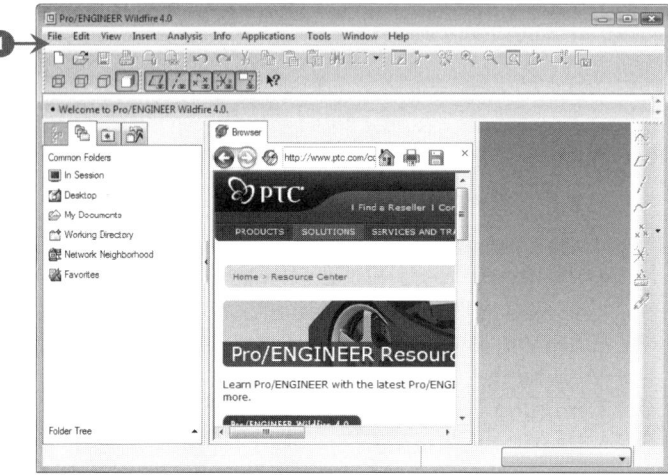

FIGURE 3.1

The **File menu** options appear (**Figure 3.2**).

2. Now, *click* the **New** option, as shown in Figure 3.2:

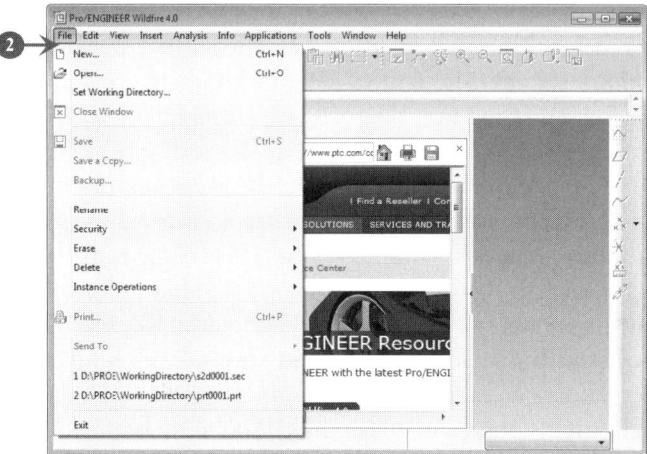

FIGURE 3.2

The **New** dialog box appears (**Figure 3.3**).

3. Now, *select* the **Sketch** radio button to enter into the **Sketch** mode, as shown in Figure 3.3.

Note: By default, the Part radio button is selected.

4. Now, *specify* a name to the sketch. To assign a name to the sketch either accepts the default name specified in the text box beside the **Name** label or specifies a new name. In our case, we have accepted the default name (Figure 3.3):

5. Next, *click* the **OK** button, as shown in Figure 3.3:

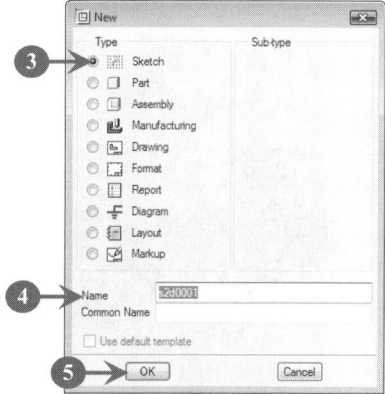

FIGURE 3.3

The Pro/ENGINEER Wildfire 4.0 window with the **Sketch** mode appears (**Figure 3.4**):

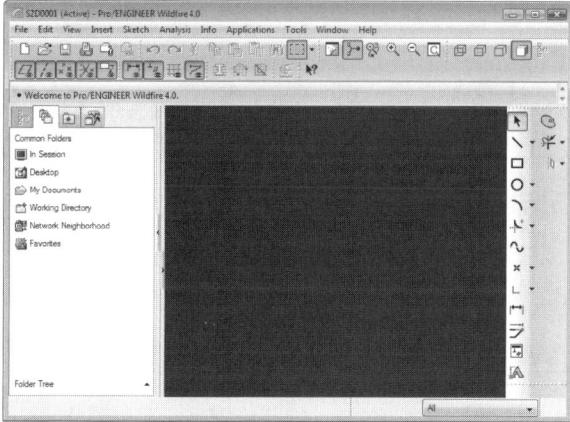

FIGURE 3.4

Now, the **Sketch** mode is ready to use. However, to draw any sketch of a feature, the **Sketcher Tools toolbar** is used. Let's explore the **Sketcher Tools toolbar** to draw sketches.

3.2 SKETCHER TOOLS TOOLBAR

The **Sketcher Tools toolbar** contains various buttons that are used to draw sketching entities such as lines, circles, and rectangles. The **Sketcher Tools toolbar** also contains buttons to dimension and constrain the sketching entities. The sketching entities when used together create a sketch of a feature. The **Sketcher Tools toolbar** is situated at the **right tool chest** in the Pro/ENGINEER Wildfire 4.0 window within the **Sketch** mode, as shown in **Figure 3.5** (see encircled part):

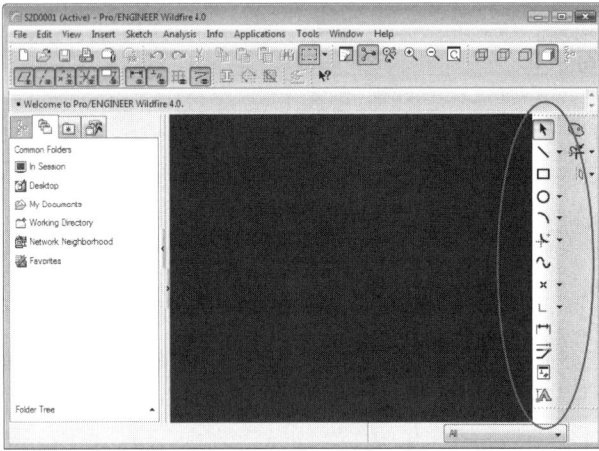

FIGURE 3.5

Note: Apart from using the **Sketcher Tools toolbar** to draw sketching entities, and dimension and constrain the sketching entities, you can use options such as Rectangle, Arc, or Circle from the **Sketch menu** in the **Menu Bar** in the Pro/ENGINEER Wildfire 4.0 window.

In this chapter, we use the **Sketcher Tools toolbar** to draw sketching entities, and dimension and constrain the sketching entities. Now, let's discuss working with the sketching entities.

3.3 WORKING WITH SKETCHING ENTITIES

Sketching entities are the basic entities used to draw a 2D sketch of a 3D feature. In Pro/ENGINEER Wildfire 4.0, a sketching entity can be drawn in two ways: first, by using the buttons available on the **Sketcher Tools toolbar**; and second, by accessing the options given in the **Sketch menu**. Sketching entities that are most commonly used to draw a 2D sketch are as follows:

- Point
- Line
- Rectangle
- Circle
- Arc
- Circular Fillet
- Elliptical Fillet
- Spline
- Text

Point

Points are geometrical entities that are used to dimension the vertices. Following are the steps to create points:

1. *Click* the **Point** (⊠▾) button on the **Sketcher Tools toolbar** (Figure 3.5).
2. Now, *click* anywhere on the **Drawing Area** to place a point.

Repeat step 2 to create more points.

> **Note:** If you create only one point, then dimensions will not appear. However, if more than one point is created, then they are automatically dimensioned. After creating all the points, *click* the middle mouse button to terminate using the Point button.

Line

In Pro/ENGINEER Wildfire 4.0, three different types of lines can be drawn. To draw three different types of lines, the following three buttons are available on the **Sketcher Tools toolbar**.

- **Line** (╲): Draws a simple line.
- **Line Tangent** (╳): Creates a tangent between two entities.
- **Centerline** (┊): Defines the axis of revolution to create a revolved feature, mirrors the sketched entities, and so on.

Note:
- A revolved feature is created by revolving a section by a specified angle around a central axis/line.
- By mirroring the sketching entities, the time to create symmetrical entities is reduced.

To access these buttons, *click* the down arrow beside the **Line** (❚▾) button in the **Sketcher Tools toolbar** (Figure 3.5). A flyout appears, as shown in **Figure 3.6**:

FIGURE 3.6

As you can see, the flyout contains three buttons to draw different types of lines. Let's perform the following steps to draw a line using the **Line** button:

1. *Click* the down arrow beside the **Line** (❚▾) button in the **Sketcher Tools toolbar** (Figure 3.5). A flyout appears (Figure 3.6).
2. *Select* the **Line** button (❚) (Figure 3.6) to create a simple line.
3. Now, *click* on any location in the **Drawing Area** (**Figure 3.7**) in the **Sketcher window** to specify the starting location of the line. A yellow rubber-band line appears that is attached to the cursor.

Note: The **Drawing Area** appears as a black screen in the **Sketcher window**.

4. Now, move the cursor over the **Drawing Area** and *click* on the desired location to end the line. A line is created, as shown in Figure 3.7:

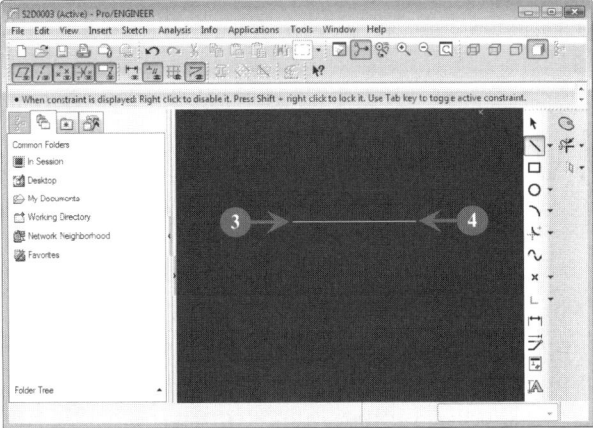

FIGURE 3.7

Another rubber band is attached to the cursor when you *click* on the second location to end the line. Now, by *clicking* on any other location, you can create another line. However, if you do not want to create another line, then *click* the middle mouse button to stop the functioning of the **Line** button. When you *click* the middle mouse button, the rubber-band line attached to the cursor disappears.

Rectangle

A rectangle in Pro/ENGINEER Wildfire 4.0 is drawn using the **Rectangle** button on the **Sketcher Tools toolbar**. The following steps are used to draw a rectangle:

1. *Click* the **Rectangle** (□) button on the **Sketcher Tools toolbar** (Figure 3.5).
2. Now, *click* any location in the **Drawing Area** (**Figure 3.8**) and drag the mouse to create a rectangle of desired size.
3. Next, *click* the left mouse button to end the rectangle. A rectangle is created, as shown in Figure 3.8:

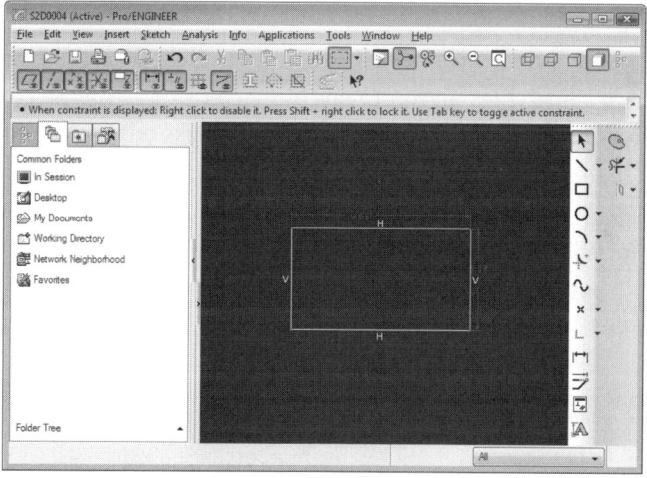

FIGURE 3.8

The **V** and **H** shown on the rectangle sides (Figure 3.8) are vertical and horizontal constraints, respectively, that represent the horizontal and vertical lines of the rectangle. The gray lines (Figure 3.8) are called weak dimensions that appear automatically and represent the horizontal and vertical dimensions of a rectangle.

Circle

In Pro/ENGINEER Wildfire 4.0, the following four buttons are available on the **Sketcher Tools toolbar** (Figure 3.5) to create a circle and one button for an ellipse.

- **Center and Point** (⊙): Creates a circle by defining the center of a circle and a point on the circle.
- **Concentric** (◉): Creates a concentric circle.
- **3 Point** (○): Creates a circle through three points.
- **3 Tangent** (○): Creates a circle that is tangent to three existing entities.
- **Ellipse** (○): Creates an ellipse.

To access these buttons, *click* the down arrow beside the **Center and Point** button ⊙˙ (Figure 3.5) in the **Sketcher Tools toolbar**. A flyout appears, as shown in **Figure 3.9**:

FIGURE 3.9

Let's perform the following steps to create a circle by using the **Center and Point** button:

1. *Click* the down arrow beside the **Center and Point** (⊙˙) button in the **Sketcher Tools toolbar** (Figure 3.5). A flyout appears, as shown in Figure 3.9.
2. Now, *select* the **Center and Point** (○) button to draw a circle (Figure 3.9).
3. Next, *click* anywhere in the **Drawing Area** to specify the center of the circle. A yellow rubber band is attached to the cursor at the center point.
4. *Drag* the cursor to get the desired circle size.
5. Now, *click* a point on the *circle*. When you *click* a point on the circle the yellow rubber band disappears and a circle is created, as shown in **Figure 3.10**:

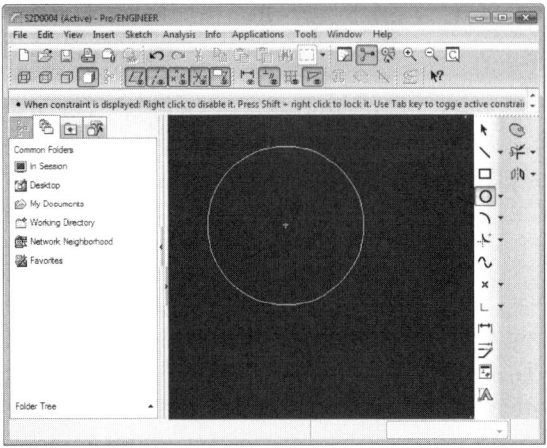

FIGURE 3.10

Continue the process to create more circles. However, if you do not want to use the **Center and Point** button further, then *click* the middle mouse button anywhere in the **Drawing Area**.

Arc

In Pro/ENGINEER Wildfire 4.0, the following five buttons are available on the **Sketcher Tools toolbar** (Figure 3.5) to draw an arc.

- **3 Point/Tangent End (⌐):** Draws an arc either by specifying three points on the **Drawing Area** or by drawing a tangent from the end of an existing entity.
- **Concentric (⌐):** Draws an arc, which is concentric to an already existing entity. The existing entity must be an arc or a circle.
- **Center and Ends (⌐):** Draws an arc by specifying a center point and an end point on the **Drawing Area**.
- **3 Tangent (⌐):** Draws an arc, which is a tangent to three selected entities.
- **Conic (⌐):** Draws a conic arc along with the selected entity.

To access these buttons, *click* the down arrow beside the **3 Point/Tangent (⌐·)** button in the **Sketcher Tools toolbar**. A flyout appears, as shown in Figure 3.11:

FIGURE 3.11

Let's perform the following steps to create an arc using the **3 Point/Tangent** button:

1. *Click* the down arrow beside the **3 Point/Tangent (⌐·)** button in the **Sketcher Tools toolbar** (Figure 3.5). A flyout appears, as shown in Figure 3.11.
2. Now, *select* the **3 Point/Tangent** button (⌐) (Figure 3.11) to draw an arc.
3. Next, *click* anywhere on the **Drawing Area** to specify the first point.
4. *Move* the cursor to size the arc.
5. Next, *click* on the **Drawing Area** to specify the second point.

6. Now, *click* anywhere on the arc to specify the third point to complete the arc creation. A sketched arc is shown in **Figure 3.12**:

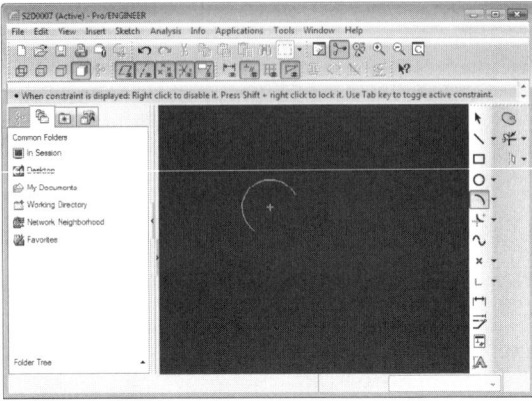

FIGURE 3.12

This is how to create an arc by specifying three points on the drawing screen.

Let's perform the following steps to create an arc, which is tangent to an already existing entity. In our case, a line already exists; therefore, we create an arc that is tangent to that line:

1. *Click* the down arrow beside the **3 Point/Tangent** ( ) button in the **Sketcher Tools toolbar** (Figure 3.5). A flyout appears, as shown in Figure 3.11.

2. Now, *select* the **3 Point/Tangent** button ( ) (Figure 3.11) to create an arc.

3. Next, *click* at the end of the line. A green circle will appear on the screen (which is known as the **Target** symbol), as pointed out in **Figure 3.13**:

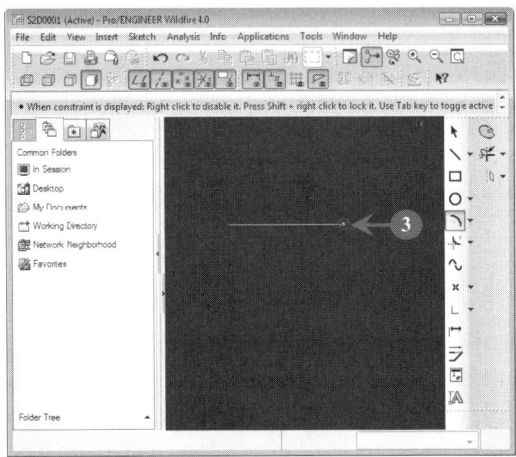

FIGURE 3.13

4. *Move* the cursor along the tangent direction to draw an arc. A yellow rubber band gets attached to the cursor.

5. *Click* anywhere in the tangent direction to complete the arc. As soon as you click, the **Target** symbol disappears. Repeat steps 3–5 to create another arc at the other end of the line; a figure is created, as shown in **Figure 3.14**:

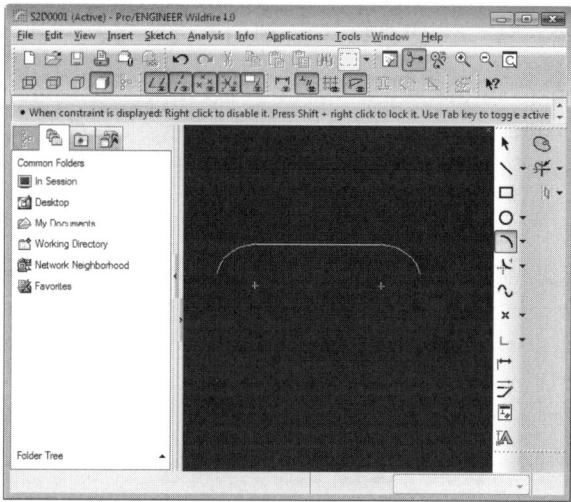

FIGURE 3.14

6. *Click* the middle mouse button to stop using the **3 Point/Tangent** button. As soon as you *click* the middle mouse button, weak dimensions appear to let you know the dimensions of the arc you have created, as shown in **Figure 3.15**:

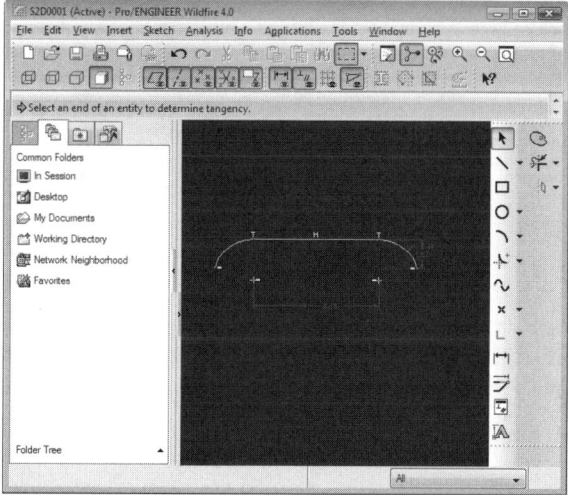

FIGURE 3.15

> **Note:** If you do not want to create an arc that is tangent to an entity, then move the cursor perpendicular to the endpoint of an entity.

Circular Fillet

A circular fillet is an arc that is used to join two entities such as two lines, two arcs, or one line-one arc. Following are the steps to create a circular fillet:

1. *Click* the down arrow beside the **Circular** ( ) button in the **Sketcher Tools toolbar** (Figure 3.5). A flyout appears, as shown in **Figure 3.16**.
2. Now, *select* the **Circular** button ( ) (Figure 3.16) to create a circular fillet, as shown in Figure 3.16:

FIGURE 3.16

In our case, we create a circular fillet to join two lines, as shown in **Figure 3.17**:

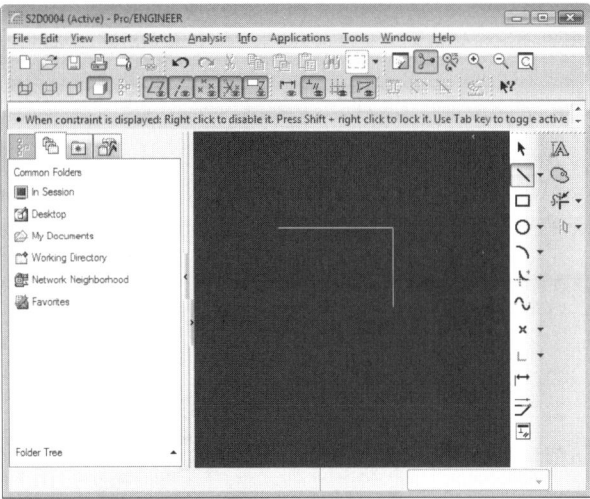

FIGURE 3.17

> **Note:** The creation of the circular fillet depends on the positions selected on the entities. For example, if you *click* anywhere on two entities, then a circular fillet is created between the clicked points. On the other hand, if you select the two corners of two entities, then a circular fillet joins the two entities from their selected corners.

3. As soon as you *click* the **Circular** button, a **Select Dialog Box** appears that prompts you to select an entity, as shown in **Figure 3.18**:

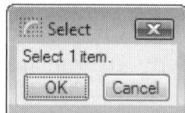

FIGURE 3.18

Now, *click* anywhere on the first entity. In our case, we clicked near the right end of the horizontal line (Figure 3.17).

4. Now, you are prompted again to select the second entity; therefore, *click* anywhere on the second entity. In our case, we clicked near the upper end of the vertical line (Figure 3.17). As soon as you *click* on the second entity, a circular fillet is automatically created, as shown in **Figure 3.19**:

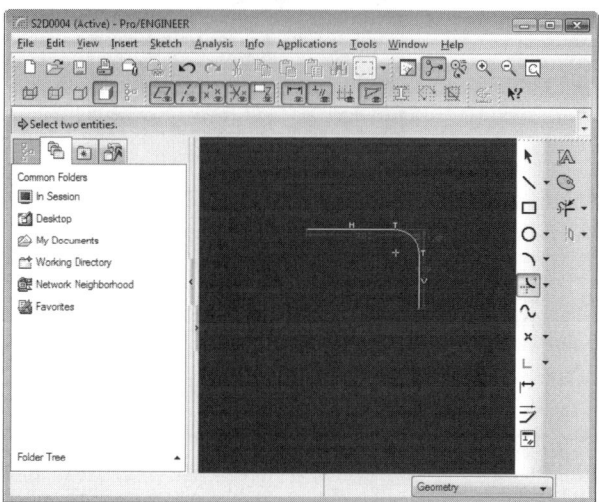

FIGURE 3.19

5. Next, *click* the **OK** button on the **Select Dialog Box** (Figure 3.18) to complete the circular fillet.

> **Note:** A circular fillet cannot be created between two parallel lines.

Elliptical Fillet

An elliptical fillet is an elliptical arc, which is used to join two entities such as two lines, one line-one arc, or two arcs. The geometry of an ellipse is controlled by dimensions in two directions: x and y directions. Therefore, when an elliptical fillet is modified, more curved geometric shapes are created. Following are the steps to create an elliptical fillet:

1. *Click* the down arrow beside the **Circular** ( ) button in the **Sketcher Tools toolbar** (Figure 3.5). A flyout appears, as shown in Figure 3.16.

2. Now, *select* the **Elliptical** ( ) button (Figure 3.16) to create an elliptical fillet. In our case, we create an elliptical fillet to join two lines, shown in **Figure 3.20**:

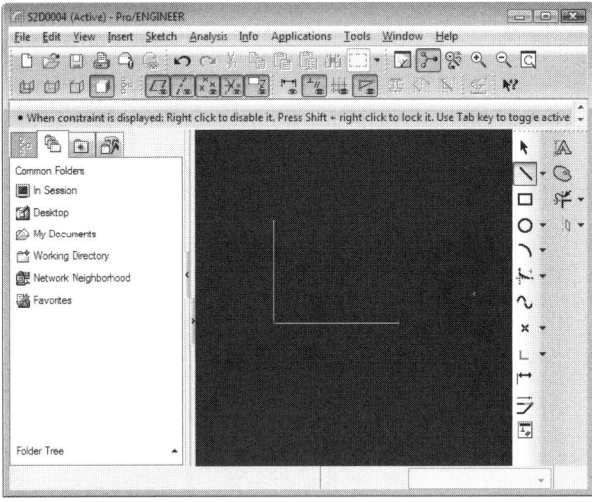

FIGURE 3.20

Note: The creation of an elliptical fillet depends on the positions selected on the entities. For example, if you *click* anywhere on two entities, then an elliptical fillet is created between the clicked points.

---

3. As soon as you *click* the **Elliptical** button, a **Select Dialog Box** (Figure 3.18) appears that prompts you to select an entity. Therefore, *click* anywhere on the first entity. In our case, we clicked near the lower end of the vertical line (Figure 3.20).

4. Now, you are prompted again to select the second entity; therefore, *click* anywhere on the second entity. In our case, we clicked near the left end of

the horizontal line (Figure 3.20). As soon as you *click* on the second entity an elliptical fillet is automatically created, as shown in **Figure 3.21**:

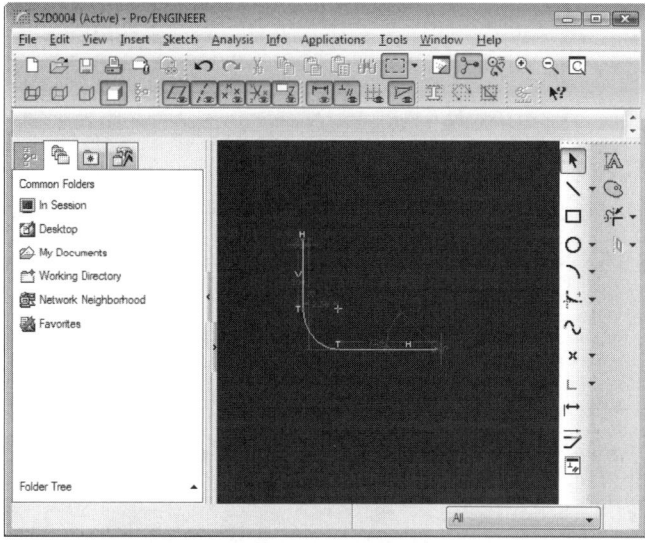

FIGURE 3.21

5.   Next, *click* the **OK** button on the **Select** **Dialog Box** to complete creating an elliptical fillet, as shown in **Figure 3.22**:

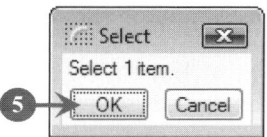

FIGURE 3.22

This is how an elliptical fillet is created. Let's move ahead by depicting another sketching entity, splines.

Spline

Splines are known as the curves that can smoothly pass through many numbers of intermediate points. Splines are very flexible; therefore, they are generally used to sketch the surface of a model such as a car, bus, or airplane. Following are the steps to sketch a spline:

1.   *Click* the **Spline** (∿) button on the **Sketcher Tools toolbar** (Figure 3.5).
2.   Now, *click* anywhere in the **Drawing Area** to select the starting point of a spline.

3. Next, *click* again on the **Drawing Area** to select the next point of the spline.

4. Repeat step 3 to select multiple points of the spline.

5. To complete the process of creating the spline *double-click* at any point in the **Drawing Area**. A spline as shown in **Figure 3.23** is created:

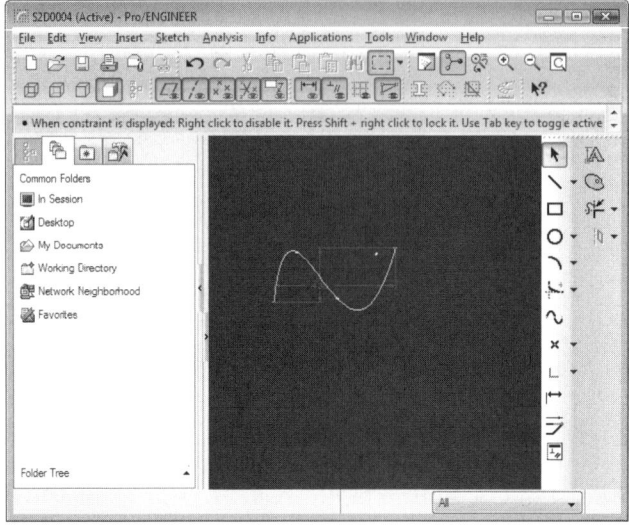

FIGURE 3.23

6. *Click* the middle mouse button to stop the functioning of the **Spline** button.

Note: The points from which a spline passes are known as interpolation points.

Text

When you create a 2D model of a 3D feature, you may need to show information such as model number and model name on the model sketch. In Pro/ENGINEER Wildfire 4.0, you can do this by using the **Text** button on the **Sketcher Tools toolbar**. Following are the steps to add text on a model sketch:

1. *Click* the **Text** (🅰) button on the **Sketcher Tools toolbar** (Figure 3.5). A message appears in the message area that prompts you to select the

start point of a line to determine text height and orientation, as shown in **Figure 3.24**:

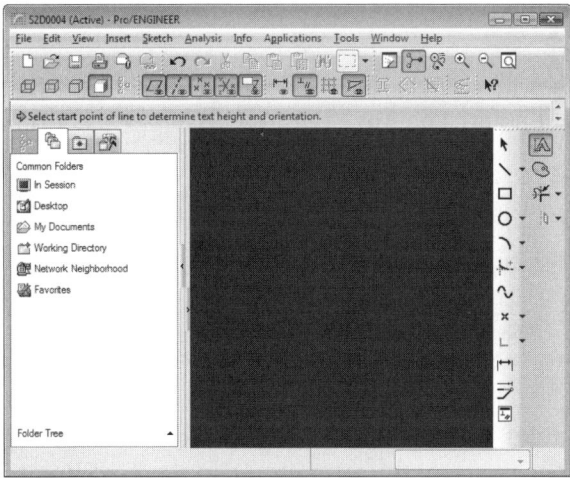

FIGURE 3.24

2. *Click* anywhere in the **Drawing Area** to select the start point. As soon as you *click* in the **Drawing Area** to select a start point, another message appears in the message area. This new message prompts you to select a second point of a line to determine text height and orientation, as shown in **Figure 3.25**:

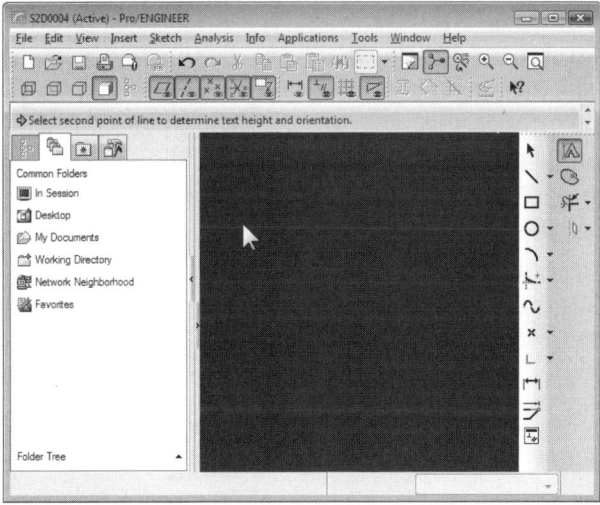

FIGURE 3.25

When you move the cursor to select the second point, a construction line appears between the start point and the endpoint. The length of the construction line determines the height of the text and the angle of the text determines the orientation of the text.

3. *Click* either in an upward or downward direction to the start point to select the second point. The **Text Dialog Box** appears (**Figure 3.26**).

Note: If you select the second point on the upward direction from the start point, then the text is written from left-to-right direction and in an upright state. However, if you want to write text in a downside state and a right-to-left state, then select the second point in the downward direction from the start point.

There are various options available in the **Font** section on **Text dialog box** to modify the appearance of the text entered in the **Text line text box**. The options available in the **Font** section on the **Text dialog box** are as follows:

- **Font:** By default, the font for the text entered in the **Text Line text box** is **font3D**. To change the font type, *click* the down arrow beside the **Font** label and select the font as per your requirement.
- **Position:** The text is positioned in horizontal and vertical directions. By default, the horizontal position of the text is left while the vertical position is bottom. To change the horizontal position, *click* the down arrow beside the **Horizontal** label in the **Position** option and select any other available position. The available horizontal text positions are **Left**, **Center**, and **Right**.
- To change the vertical position, *click* the down arrow beside the **Vertical** label in the **Position** option and select any other available position. The available horizontal text positions are **Bottom**, **Middle**, and **Top**.
- **Aspect Ratio:** The aspect ratio is modified by moving the slider in a left or right direction.
- **Slant Angle:** The slant angle can be modified by moving the slider in a left or right direction.

4. *Enter* some text in the **Text line text box**. By default, the **Text Line text box** is empty. In our case, we have entered "Text Button" in this text box (Figure 3.26).

> **Note:** You can enter the text in the **Text Line text box** up to a length of 79 characters. To add any symbol in this text box, *click* the **Text Symbol** button on the **Text dialog box**. The **Text Symbol dialog box** appears, select the symbol to insert in the text box, and *click* the **Close** button to close the **Text Symbol dialog box**.

    5.   Now, *click* the **Ok** button to finish, as shown in Figure 3.26:

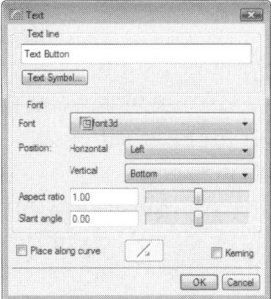

FIGURE 3.26

The **Text dialog box** is closed. The text entered in the **Text Line text box** appears on the **Drawing Area**, as shown in **Figure 3.27**:

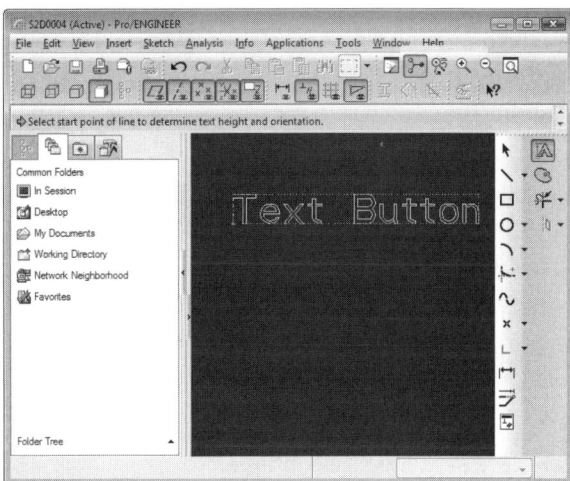

FIGURE 3.27

Note: You can *click* the **Cancel** button in the **Text dialog box** to exit from it without doing any action.

This discussion was about drawing the sketch entities. Now, the next step in sketch creation is dimensioning the sketch entities.

3.4 WORKING WITH DIMENSIONS

Whenever a sketching entity is created, some dimensions are automatically applied on that entity. These dimensions are known as weak dimensions and are gray in color. These dimensions are called weak dimensions because they are generated and removed automatically without any confirmation.

Apart from the weak dimensions, you can apply your own dimensions on the sketched entities. Dimensions that you apply on sketched entities are known as strong dimensions that are mainly used to control their size. When you apply strong dimensions on any sketched entity, the weak dimensions are automatically deleted.

Dimensions can be added on sketched entities in two ways: first, by selecting the **Dimensions** option in the **Sketch menu** in the Pro/ENGINEER Wildfire 4.0 window; and second, by using the **Normal** button available on the **Sketcher Tools toolbar** in the Pro/ENGINEER Wildfire 4.0 window. In this section, we see dimensioning the sketched entities using the **Normal** button by following the steps given here:

1. *Click* the Normal (⊢) button on the **Sketcher Tools toolbar** (Figure 3.5). A **Select dialog box**, as shown in **Figure 3.28**, appears that prompts you to select an entity:

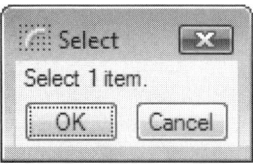

FIGURE 3.28

In our case, we have sketched a rectangle.

2. Now, *select* one of the horizontal lines of the rectangle to apply dimensions on it. As soon as you *select* the horizontal line, the color of the line is changed to red from yellow (**Figure 3.29**).

3. Next, *move* the cursor on the selected line and *click* the middle mouse button to apply dimensions on it (Figure 3.29):

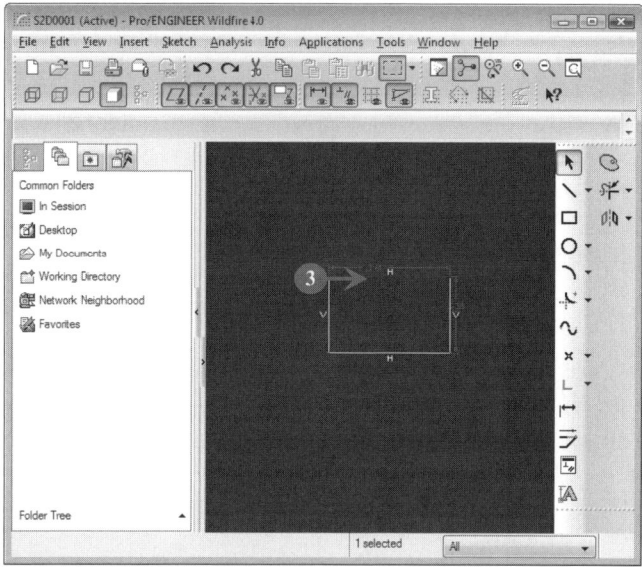

FIGURE 3.29

As soon as you *click* the middle mouse button, dimension is applied on the selected line of the rectangle, as shown in **Figure 3.30**:

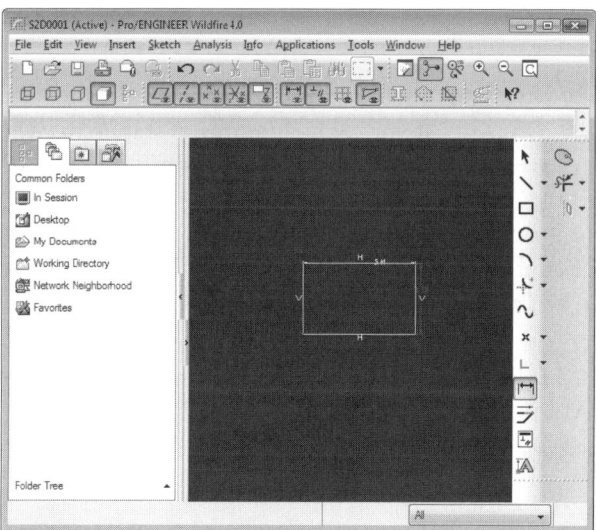

FIGURE 3.30

Note: The color of the horizontal line is changed to yellow once again.

4. Now, *click* the **OK** button to stop using the **Normal** button, as shown in **Figure 3.31**:

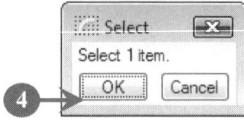

FIGURE 3.31

This is how the **Normal** button is used to dimension the sketched entities.

Next, we discuss the different types of dimensions to apply on the basic sketched entities under the following subheadings:

- Linear
- Radial
- Angular
- Diameter

Let's start depicting the **Linear** dimension.

Linear

Linear dimensioning calculates the length of a line, the distance between two parallel lines, the distance between a point and a line, and the distance between two points. Let's perform the following steps to apply **Linear** dimensioning on a line:

1. *Click* the **Normal** (⊡) button on the **Sketcher Tools toolbar** (Figure 3.5). A **Select dialog box** appears, as shown in Figure 3.31, that prompts you to select an entity. In our case, we have sketched a line.
2. Now, *select* the line to apply dimensions on it. As soon as you *select* the line, the color of the line is changed to red from yellow (**Figure 3.32**).
3. Next, *move* the cursor on the selected line and *click* the middle mouse button to apply dimensions on it (Figure 3.32):

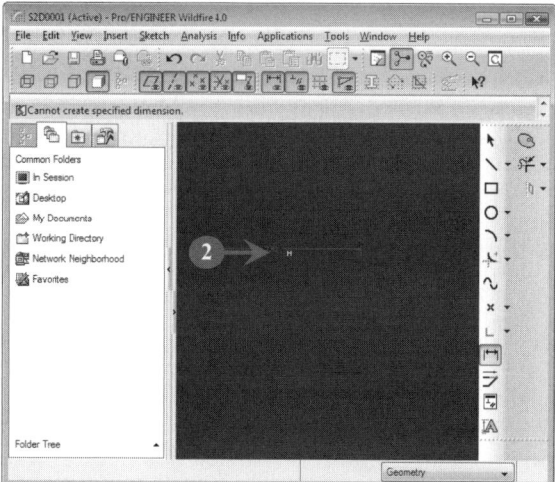

FIGURE 3.32

As soon as you *click* the middle mouse button, dimension is applied on the line, as shown in **Figure 3.33**:

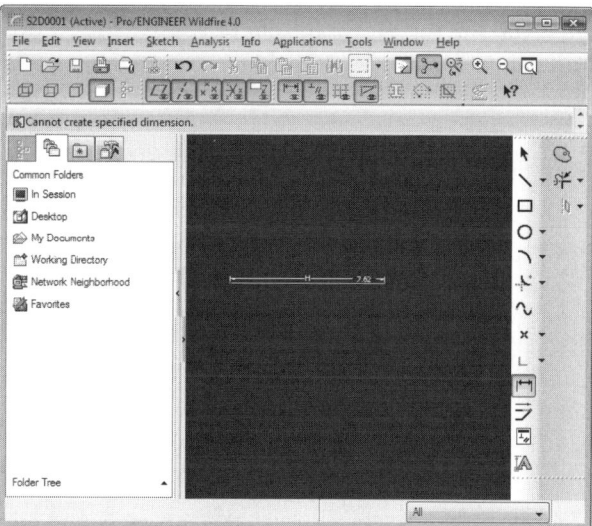

FIGURE 3.33

The color of the line gets changed to yellow once again.

Note:

- The location on the line where the dimension will appear depends on the point where you *click* the middle mouse button.
- A center line cannot be dimensioned because it is infinite.

4. Now, *click* the **OK** button to stop using the **Normal** button (Figure 3.31).

Radial

Radial dimensioning is used to calculate the radius of an arc or a circle. In this section, we see the process to apply **Radial** dimension on a circle. Following are the steps to apply **Radial** dimension on a circle:

1. *Click* the **Normal** (⊨) button on the **Sketcher Tools toolbar** (Figure 3.5). A **Select dialog box**, as shown in Figure 3.31, appears that prompts you to *select* an entity. In our case, we use circle as an entity.

2. Now, *select* the circle to apply dimensions on it. As soon as you select the circle, weak dimension is applied on the circle (**Figure 3.34**).

3. Next, *move* the cursor on the circle and *click* the middle mouse button to apply dimensions on it (Figure 3.34):

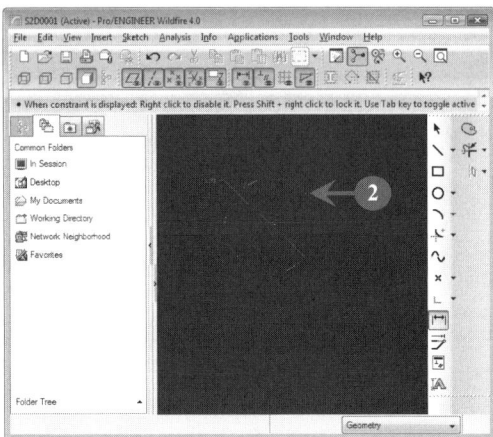

FIGURE 3.34

As soon as you *click* the middle mouse button, dimension is applied on the circle, as shown in **Figure 3.35**:

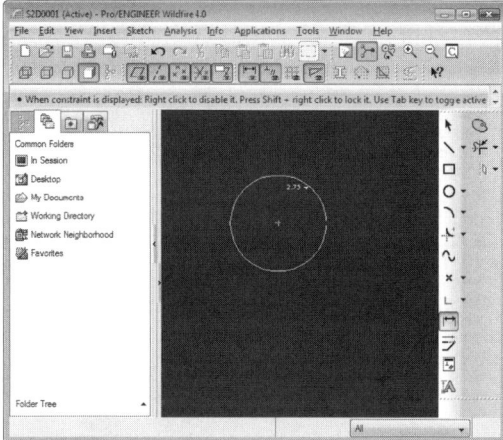

FIGURE 3.35

The radius of the circle appears in white, as shown in Figure 3.35. In our case, radius of the circle is 2.75.

Note: The location on the circle where the radius will appear depends on the point where you *click* the middle mouse button.

4.   Now, *click* the **OK** button to stop using the **Normal** button (Figure 3.31).

Angular

Angular dimensioning is used to calculate the angle of an arc or the angle between two lines. In this section, we see the process to calculate an angle between two lines. Following are the steps to apply **Angular** dimensioning on a circle:

1.   *Click* the **Normal** (⊟) button on the **Sketcher Tools toolbar** (Figure 3.5). A **Select dialog box** appears, as shown in Figure 3.31, that prompts you to select an entity. In our case, we use two lines, shown in **Figure 3.36**.

2. Now, *select* the horizontal line (Figure 3.36). As soon as you *select* the horizontal line, the **Select dialog box** appears again that prompts you to select a line. The prompt *select a line* here means to select the second line (Figure 3.31).

3. Next, *select* the second line, as shown in Figure 3.36:

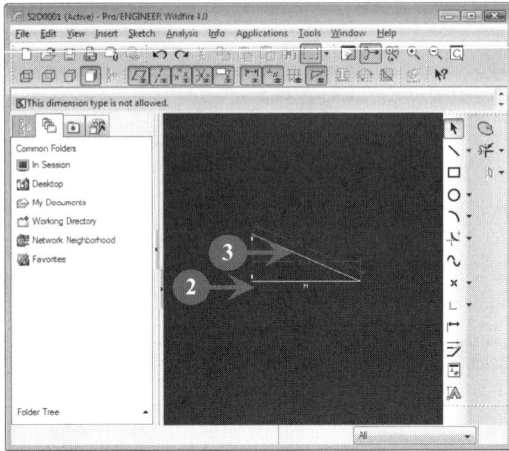

FIGURE 3.36

4. Now, *click* the middle mouse button on any part of the line. **Angular** dimension is applied on the lines, as shown in **Figure 3.37**:

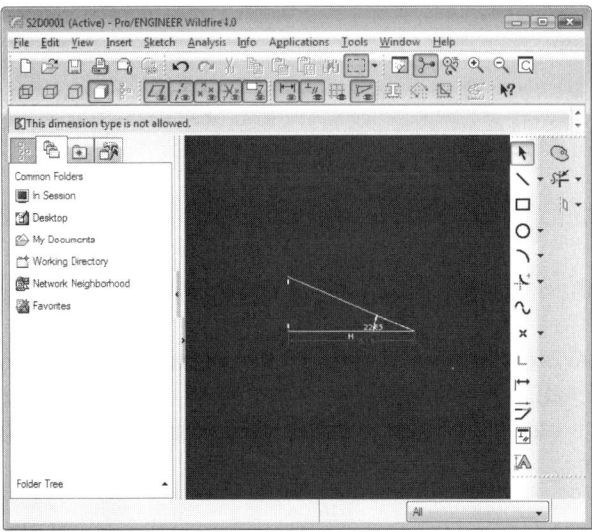

FIGURE 3.37

The angle between two lines appears in white, as shown in Figure 3.37. In our case, the angle is 22.85°.

> **Note:** Depending on the location of the dimension, the angle can be acute or obtuse.

5. Finally, *click* the **OK** button to stop using the **Normal** button (Figure 3.31).

Diameter

Diameter dimensioning is used to calculate the diameter of a circle or on an arc. In this section, we discuss the process to apply **Diameter** dimensioning on a circle. Follow the steps given here to apply **Diameter** dimensioning on a circle:

1. *Click* the **Normal** (⌨) button on the **Sketcher Tools toolbar** (Figure 3.5). A **Select dialog box**, as shown in Figure 3.31, appears that prompts you to select an entity. In our case, we use a circle as an entity.
2. Now, *double-click* the circle.

> **Note:** By *double-clicking*, the diameter is applied; but if you only single-click, the radius of the circle is applied.

3. Next, *click* the middle mouse button to apply dimension on the circle. **Diameter** dimensioning is applied on the circle, as shown in **Figure 3.38**:

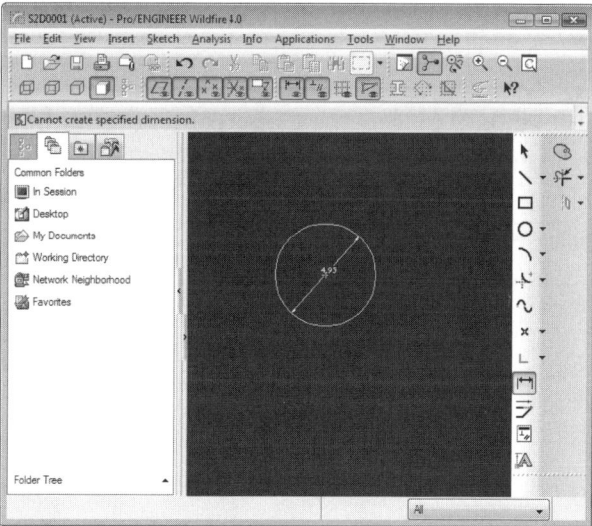

FIGURE 3.38

The diameter of the circle appears in white, as shown in Figure 3.38. In our case, the diameter is 4.95.

4.   Now, *click* the **OK** button to stop using the **Normal** button (Figure 3.31).

Note: Apart from dimensioning line, arc, and circle, you can also dimension splines. To dimension splines, use the **Coordinate System** button available on the **Sketcher Tools toolbar**.

This is how entities are dimensioned.Next, we study applying constraints on sketching entities, because an entity in Pro/ENGINEER Wildfire 4.0 cannot be completed until it is fully dimensioned and constrained.

3.5 APPLYING CONSTRAINTS

Constraints are applied on entities in a 2D sketch to make them accurate in respect to the position of other entities. Constraints are applied on an entity by using the buttons available in the **Constraints dialog box**.

Note: The process to access the **Constraints dialog box** is discussed ahead in this section.

Let's now see **Figure 3.39** to view the **Constraints dialog box**:

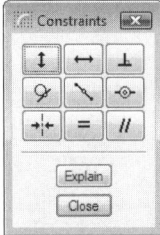

FIGURE 3.39

Table 3.1 lists the buttons available in the **Constraints dialog box**:

Button	Function
↕	Causes a selected line or two vertices to be vertical.
↔	Causes a selected line or two vertices to be horizontal.
⊥	Causes the two entities to be perpendicular.
◌	Causes the two entities to be tangent.
↘	Places a selected point or vertex on the middle of a line.
-◉-	Makes the two selected entities collinear. This button also makes two points coincident.
⊣⊦	Causes two points or vertices to be symmetric about a centerline.
=	Makes the two selected entities such as lines, arcs, circles, or ellipses become equal in dimensions. If you apply this constraint on two lines, then those lines become equal in their length. However, if you apply this constraint on arcs, circles, or ellipses, then these entities become equal with respect to their radii.
//	Causes two lines to be parallel.
Explain	Provides information about the constraints applied to a sketch.
Close	Closes the Constraints **dialog box**.

Table 3.1   Constraints dialog box

Now, you are acquainted with the buttons available on the **Constraints dialog box**; therefore, next we can study the process to create constraints on a sketch. In this section, we discuss the use of the (=) button. Follow the steps given here to apply **Constraints** on a sketch:

1.   *Click* the **Constrain** ( ) button on the **Sketcher Tools toolbar** (Figure 3.5). The **Constraints dialog box** appears (Figure 3.39).
2.   Now, *click* the (=) button in the **Constraints dialog box**. A **Select dialog box** (Figure 3.31) appears that prompts you to select an entity. In this section, we apply equal length constraints on two lines of unequal length (**Figure 3.40**).
3.   Next, *select* the left vertical line (Figure 3.40).

4. Now, *select* the right vertical line, as shown in Figure 3.40:

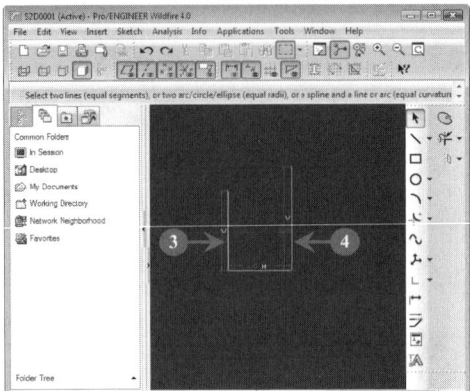

FIGURE 3.40

As soon as you *select* the right vertical line both of the lines become equal, as shown in **Figure 3.41**:

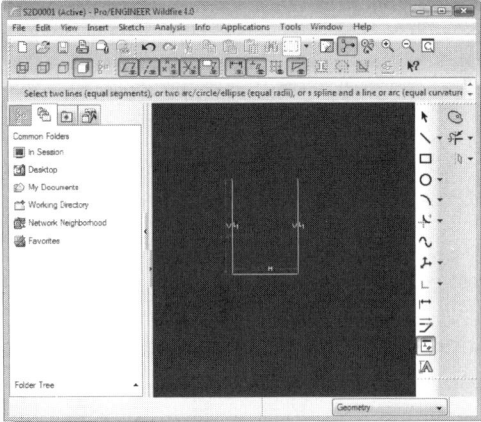

FIGURE 3.41

5. Next, *click* the **OK** button to close the **Select dialog box** (Figure 3.31).
6. Finally, *click* the **Close** button on the **Constraints dialog box** (Figure 3.39) to close it.

This is how **Constraints** are applied on a sketch. Next, we discuss modifying dimensions.

3.6 MODIFYING DIMENSIONS

Once you have completed a sketch, you may need to **Modify** the dimensions of one or more entities in that sketch to make those entities accurate with other entities. To

Modify the dimensions of the entities in a sketch, the following four options are available in Pro/ENGINEER Wildfire 4.0:

- Modify Button
- Modify Option in the Edit Menu
- Double-Clicking a Dimension
- Modifying Dimension Dynamically

Let's start with the **Modify** button.

Modify Button

The **Modify** (⧎) button is available on the **Sketcher Tools toolbar** (Figure 3.5) and is the most commonly used option to modify dimensions. Following are the steps that let you modify dimensions using the **Modify** button:

1. *Click* a dimension in a sketch that you want to modify. When you *click* a dimension, its color becomes red. In our case, we are modifying the dimensions of a line, as shown in **Figure 3.42**:

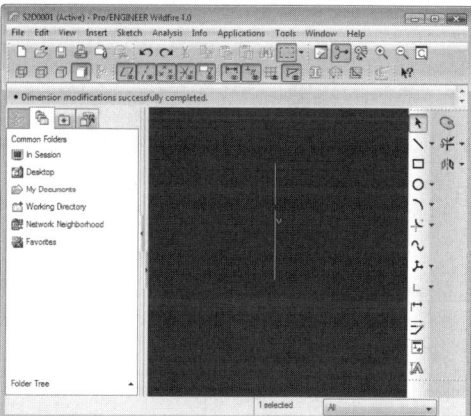

FIGURE 3.42

Note: Hold the CTRL key if you want to select more than one dimension to modify.

2. Now, *click* the **Modify** (⧎) button on the **Sketcher Tools toolbar** (Figure 3.5). The **Modify Dimensions dialog box** appears (**Figure 3.43**).

3. *Modify* the dimension value appearing in the text box (Figure 3.43) by entering the new dimension value. The value appearing in the text box (Figure 3.43) is the current dimension of the entity you want to modify. In our case, we entered 2.78872346793 as the new value.

4. *Click* the **Build** (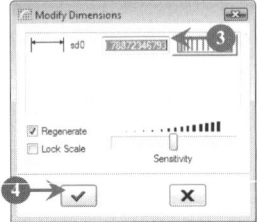) button to reflect the changes in the **Drawing Area**, as shown in Figure 3.43:

FIGURE 3.43

The dimension of the selected line is changed, as shown in **Figure 3.44**:

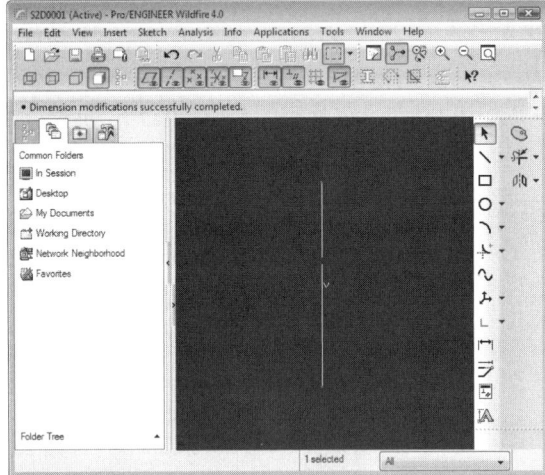

FIGURE 3.44

Now, if you compare the line dimension appearing in Figure 3.42 with the dimension appearing in Figure 3.44, then you notice that the length of the line in Figure 3.44 is increased. This is because the new dimension value, 2.78872346793, is more than the older value, 1.79.

Modify Option in the Edit Menu

Dimensions in a sketch can also be modified using the **Modify** option in the **Edit menu**. To modify dimensions using the **Modify** options in the **Edit menu,** follow the steps given here:

1. *Click* **Edit > Modify...** in the **Menu Bar**. A check mark appears on the left of the **Modify** option in the **Edit menu**, which means that the **Modify** option can be used now.

2. Next, *select* a dimension in a sketch to modify. As soon as you select a dimension, the **Modify Dimensions dialog box** appears (Figure 3.43).

3. *Enter* the new dimensions to modify the current dimensions in the text box on the **Modify Dimensions dialog box**.

4. *Click* the **Build** (✔) button to reflect the changes in the **Drawing Area**.

Double-Clicking a Dimension

Another way to modify a dimension is to **Double-Click** a dimension. Follow the steps given here to modify a dimension by **Double-Clicking**:

1. *Double-click* a dimension in a sketch. A text box appears showing you the current dimension.

2. *Edit* the current dimension.

3. Now, either *press* the **ENTER** key or *click* the middle mouse button on the **Drawing Area**. Changes are reflected in the **Drawing Area**.

Modifying Dimension Dynamically

The dimension of an entity can also be modified by dragging the dimension. Following are the steps that help you to modify a dimension by dragging it:

1. *Move* the cursor on the entity whose dimension you have to modify. When you move the cursor on an entity, its color becomes cyan.

2. Now, *place* the cursor on the end of an entity and *drag* it. When you drag the entity, along with the entity its dimension also changes. The changed dimension appears as a weak dimension along with the entity.

Next, we move ahead by depicting the process to trim entities.

3.7 TRIMMING ENTITIES

Whenever a sketch is created, some unwanted areas may also be created in the various entities in that sketch. To remove those areas, you need to trim the entities. In Pro/ENGINEER Wildfire 4.0, the following three buttons are available in the **Sketcher Tools toolbar** that can be used to trim entities:

- Delete Segment Button
- Corner Button
- Divide Button

Let's discuss the process of trimming entities using these three buttons.

Delete Segment Button

The **Delete Segment** () button deletes the selected entity. Follow the steps given here to use the **Delete Segment** button to trim an entity in a sketch:

1. *Click* the down arrow beside the **Delete Segment** () button in the **Sketcher Tools toolbar** (Figure 3.5). A flyout appears (**Figure 3.45**).

2. Now, *select* the **Delete Segment** () button, as shown in Figure 3.45:

FIGURE 3.45

3. Next, *move* the cursor over the entity you want to trim in the sketch. In our case, the sketch is shown in **Figure 3.46**:

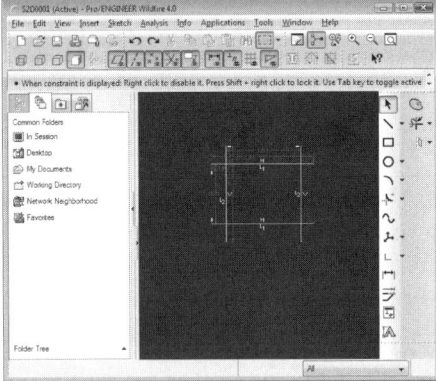

FIGURE 3.46

When you move the cursor on an entity, the color of that entity becomes cyan.

4. *Click* the left mouse button. The entity is deleted. In our case, the sketch after trimming an entity is shown in **Figure 3.47**:

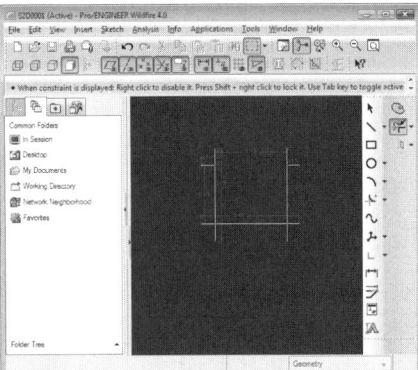

FIGURE 3.47

Corner Button

The **Corner** button in Pro/ENGINEER Wildfire 4.0 trims two entities at their corner. When using the **Corner** button to trim entities, you must remember that only the selected portion of an entity is retained and the other is trimmed. Follow the steps given here to trim entities using the **Corner** button:

1. *Click* the down arrow beside the **Delete Segment** (⛐ ▾) button in the **Sketcher Tools toolbar** (Figure 3.5). A flyout appears, as shown in Figure 3.45.
2. Now, *select* the **Corner** (┼) button (Figure 3.45). The **Select dialog box** appears (Figure 3.31).
3. *Select* the first entity; you want to trim in the sketch (Figure 3.46). In our case, we select the left vertical line in the sketch (Figure 3.46). Now, you prompt again to select the second entity.
4. *Select* the second entity in the sketch (Figure 3.46). In our case, we selected the upper horizontal entity in the sketch (Figure 3.46).

As soon as you *select* the second entity, both entities are trimmed, as shown in **Figure 3.48**:

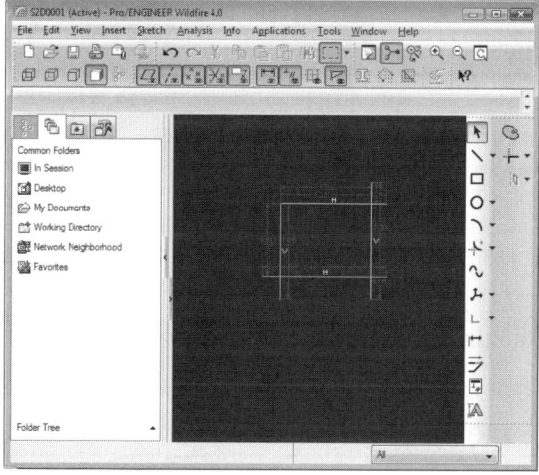

FIGURE 3.48

5. Now, *click* the **OK** button (Figure 3.31) on the **Select dialog box** to stop using the **Corner** button.

Divide Button

The **Divide** button divides an entity in many parts depending on the points specified in that entity. Follow the steps given here to use the **Divide** button:

1. *Click* the down arrow beside the **Delete Segment** (📉 ▾) button in the **Sketcher Tools toolbar** (Figure 3.5). A flyout appears, as shown in Figure 3.45.

2. Now, *select* the **Divide** (🖋) button (Figure 3.45).

3. *Move* the cursor over the entity (Figure 3.46).

4. Now, *click* on the point over the entity where you want to divide. In our case, we *click* in the middle of the upper horizontal line (Figure 3.46). The selected entity is divided in two entities, as shown in **Figure 3.49**:

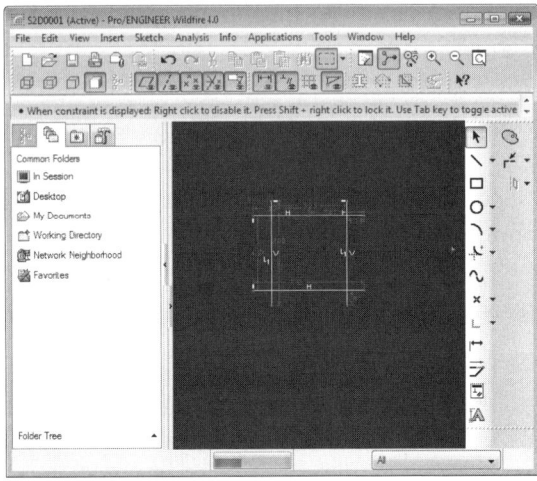

FIGURE 3.49

You can see a very small blue dot in the middle of the upper horizontal line, as shown in Figure 3.49, which means the line is divided into two parts. Now, this line can be treated as two separate entities. Apart from dividing a line, the **Divide** button can also be used to divide entities such as an arc or a circle.

Next, we discuss the process of mirroring entities.

3.8 MIRRORING ENTITIES

Mirroring is the process of creating and dimensioning the symmetrical entities. Mirroring saves you time by creating symmetrical entities in just a few steps. How mirroring saves time is discussed ahead in this section. In Pro/Engineer Wildfire 4.0, mirroring can be done in two ways: first, through using the **Mirror** button on the **Sketcher Tools toolbar**; and second, by accessing the **Mirror** option in the **Edit menu**. In this chapter, we use the **Mirror** button on the **Sketcher Tools toolbar** to mirror entities. To mirror entities, you need a sketch; in our case, the sketch is shown in **Figure 3.50**:

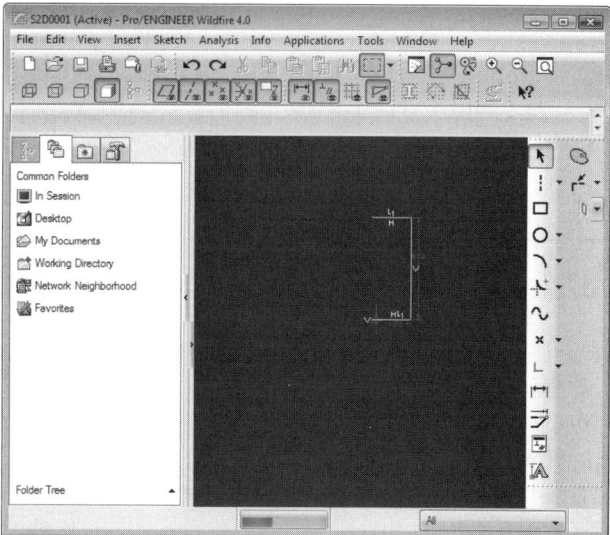

FIGURE 3.50

Let's perform the following steps to mirror the entities:

1. *Click* the down arrow beside the **Line** (✎) button in the **Sketcher Tools toolbar** (Figure 3.5). A flyout appears, as shown in Figure 3.6.

2. *Select* the **Centerline** (┊) button (Figure 3.6). The **Centerline** button is selected to create an axis of revolution along with the entity to be mirrored.

3. Now, *click* on the **Drawing Area** to create a centerline around the entity you want to mirror, as shown in **Figure 3.51**:

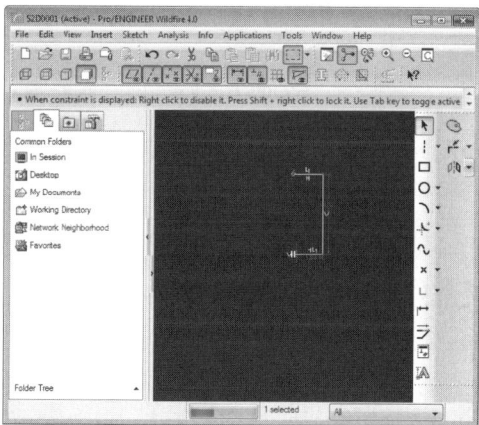

FIGURE 3.51

The infinite line (Figure 3.51) is called the centerline.

4.   Next, *select* an entity to mirror. If you want to select multiple entities, hold the **CTRL** key and *select* **Entities**. In our case, we select all the entities appearing in Figure 3.51.

When you select an entity, the **Mirror** ( ) button is enabled.

Note: By default, the **Mirror** button is disabled.

5.   Now, *click* the **Mirror** button on the **Sketcher Tools toolbar**. As soon as you *click* the **Mirror** button, the **Select dialog box** (Figure 3.31), along with a message on the **Message** area, appears that prompts you to select the centerline.

6.   Next, *click* the centerline. The entities you have selected are mirrored, as shown in **Figure 3.52**:

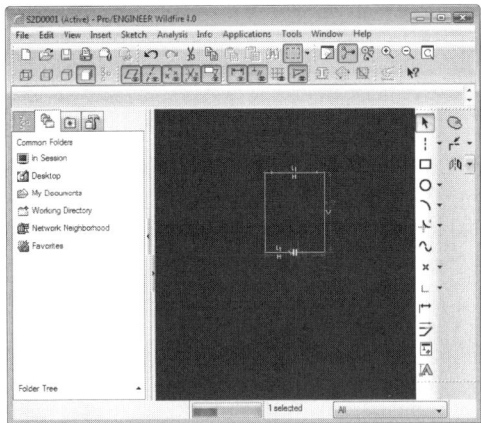

FIGURE 3.52

In this way, entities are mirrored. Let's next discuss the process to scale and rotate the entities.

3.9 SCALING AND ROTATING ENTITIES

While drawing a 2D sketch, you may need to scale and rotate one or more entities to get the desired sketch. In Pro/Engineer Wildfire 4.0, entities can be scaled and rotated in two ways: first, through using the **Scale and Rotate** button available on the **Sketcher Tools toolbar**; and second, by accessing the **Scale and Rotate** option in the **Edit menu**. In this chapter, we use the **Scale and Rotate** button available on the **Sketcher**

Tools toolbar to scale and rotate the entities. To scale and rotate the entities you need a sketch, and in our case the sketch is shown in Figure 3.51.

Let's now follow the steps given here to scale and rotate the entities:

1. *Select* the entity you want to scale and rotate (Figure 3.51).
2. *Click* the down arrow beside the **Mirror** () button in the **Sketcher Tools toolbar** (Figure 3.5). A flyout appears (**Figure 3.53**).
3. *Click* the **Scale and Rotate** (⊚) button, as shown in Figure 3.53:

FIGURE 3.53

As soon as you *click* the **Scale and Rotate** button, the selected entity is enclosed in a boundary, as shown in **Figure 3.54**:

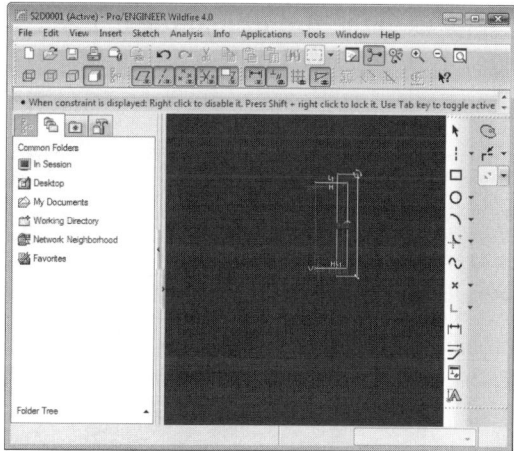

FIGURE 3.54

Apart from this, the **Scale and Rotate dialog box** also appears, as shown in **Figure 3.55**:

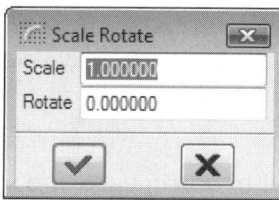

FIGURE 3.55

Now, you have the following two approaches to scale and rotate the selected entity:

- First, modify the values in the text boxes beside the **Scale and Rotate** labels in the **Scale and Rotate dialog box**. After modifying the values, *click* the **Accept** (✔) button to accept the changes. As you *click* the **Accept** button, changes are reflected in the **Drawing Area**.
- Second, use the various handles available in Pro/ENGINEER Wildfire 4.0 to dynamically scale and rotate the selected entity. The following handles are available on the boundary of the selected entity to dynamically scale and rotate the selected entity:
  - **Rotate Handle:** Used to dynamically rotate the selected entity
  - **Base-Point Handle:** Used to move the selected entity from one place to another in the **Drawing Area**
  - **Scale Handle:** Used to scale the selected entity

In our case, we have used the **Rotate handle** to rotate the selected entity, and the resultant figure is shown in **Figure 3.56**:

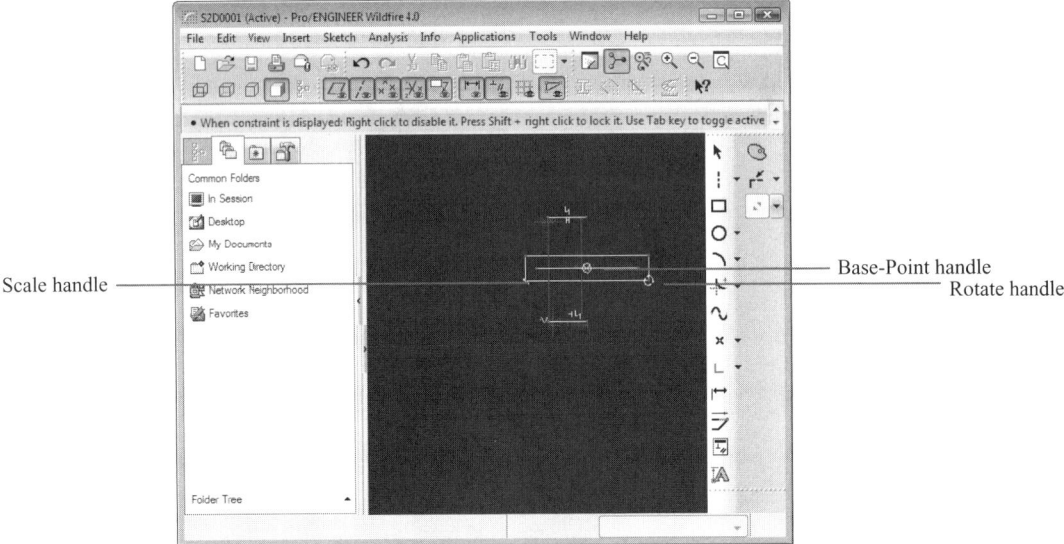

FIGURE 3.56

4. Now, *click* the **Accept** button in the **Scale and Rotate** (Figure 3.55) **dialog box** to accept the changes. Changes are reflected in the **Drawing Area**, as shown in **Figure 3.57**:

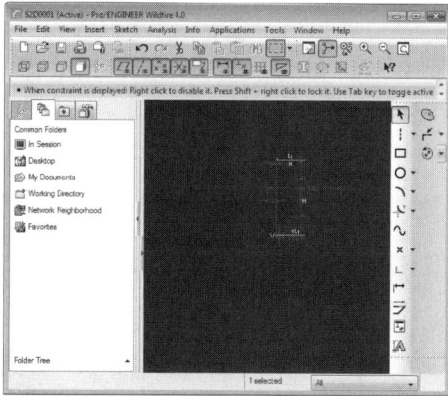

FIGURE 3.57

With this, we complete the discussion on **Scaling and Rotating** the entities. Next, we discuss the process to import an already created sketch.

3.10 IMPORTING 2D DRAWINGS

Sometimes, you may need to import an already created sketch to create another 2D sketch. For example, you have created the design of a pen a few days ago, but now you want to redesign the pen with a different cap design. Now, you can import the preexisting sketch of the pen, change the cap design, and the new pen is ready. This saves you time while creating new sketches from the preexisting sketches. The **Sketches/Drawings** created in **Sketcher** mode are saved with the .sec extension so that they can be used to create other 2D sketches. The process to save a file is depicted in Chapter 2, "Exploring the User Interface." However, to use an already created 2D sketch, you have to import that sketch in the **Sketch** mode. Following are the steps to import an already created sketch:

1. *Click* the **Sketch>Data from File>File System**, as shown in **Figure 3.58**:

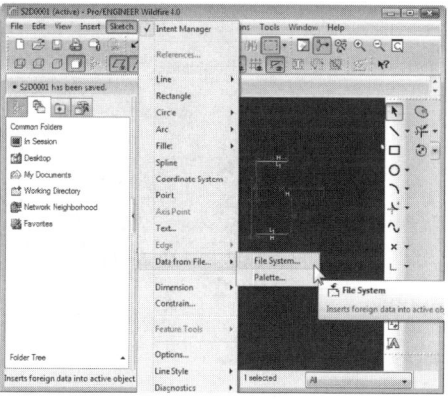

FIGURE 3.58

The **Open dialog box** appears (**Figure 3.59**).

2. Now, *select* the file you want to open (Figure 3.59). In our case, we have selected the s2d0001.sec file.

3. Next, *click* the **Open** button, as shown in Figure 3.59:

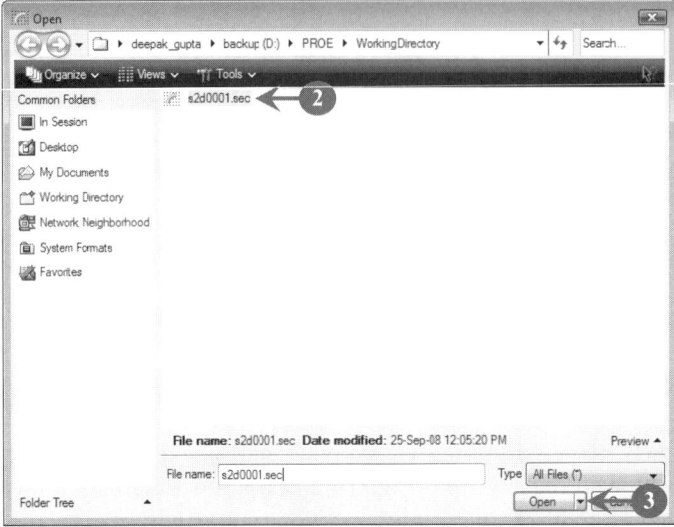

FIGURE 3.59

The selected file is opened in the **Drawing Area**, as shown in **Figure 3.60**:

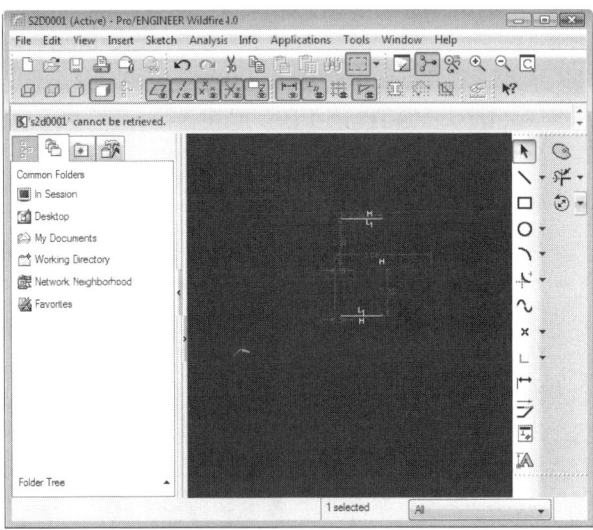

FIGURE 3.60

Now, you can use this file to create another 2D sketch. Let's now discuss the process to delete entities from a sketch.

3.11 DELETING ENTITIES

Whenever creating a sketch in Pro/ENGINEER Wildfire 4.0, you may need to delete some extra entities to complete the sketch. In Pro/ENGINEER Wildfire 4.0, you have the following two ways to delete entities:

- First, select the entity you want to delete and then press the **DELETE** key on your keyboard.

Note: To delete more than one entity, select multiple entities by holding the **CTRL** key.

- Second, use the **Delete** option in the **Edit menu**.

In this chapter, we use the **Delete** option to delete the entities. For deleting entities, we use the sketch shown in Figure 3.59. Let's now follow the steps given here to delete the entities from a sketch:

1. *Select* an entity (Figure 3.60) that you want to delete.
2. Now, *click* the **Edit > Delete** option, as shown in **Figure 3.61**:

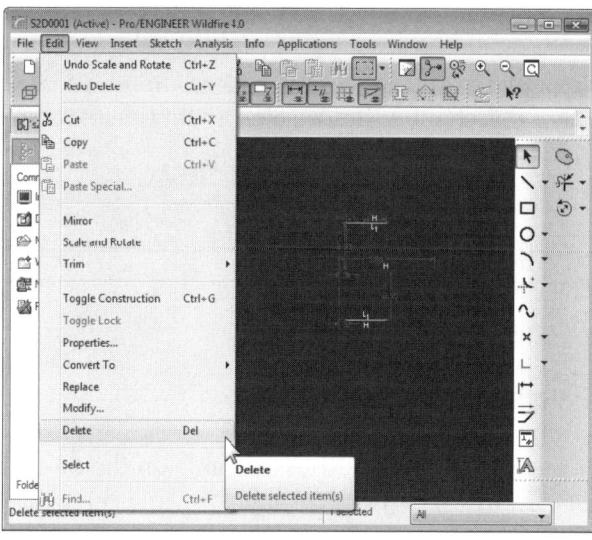

FIGURE 3.61

The selected entity is automatically deleted, as shown in **Figure 3.62**:

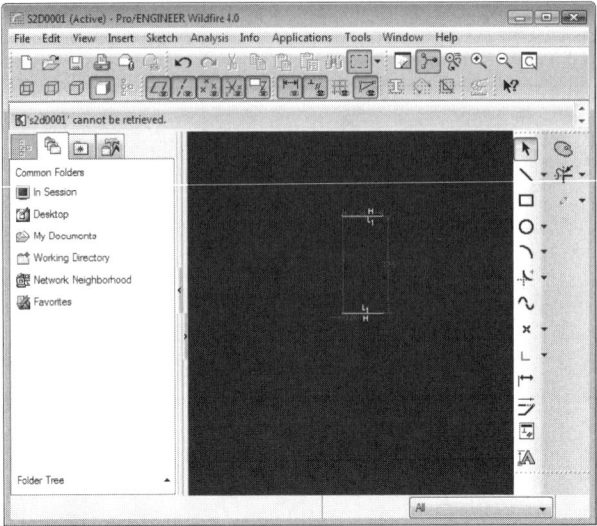

FIGURE 3.62

> **Note:** The **Delete** option is enabled when you select an entity. By default, the **Delete** option is disabled.

Here, we have completed the process to delete the entities. Apart from this, we have reached the end of this chapter. Before closing this chapter, let's have a look at the topics covered in this chapter.

SUMMARY

In this chapter, you learned about:

- The process to start the **Sketch** mode
- The **Sketcher Tools toolbar**
- The processes to use the buttons available on the **Sketcher Tools toolbar** to sketch various entities such as line, arc, and rectangle
- The process of applying **Dimensions** on sketching entities
- The process of applying **Constraints** on sketching entities

- Various options to **Modify Dimensions**
- The buttons available on the **Sketcher Tools toolbar** to trim the entities
- The processes to **Mirror**, **Scale**, and **Rotate** the entities
- The process to import already created sketch files
- The process to delete entities from a sketch

EXPLORING PRO/ENGINEER WILDFIRE 4.0 PART MODE

4

In This Section

◊ Starting the Part Mode
◊ Understanding Datums
◊ Creating an Extruded Feature
◊ Inducing the Hole Feature
◊ Creating a Cut Feature
◊ Inducing the Round Feature
◊ Inducing the Chamfer Feature
◊ Inducing a Shell Feature
◊ Inducing a Revolved Feature
◊ Working with the Pattern Feature
◊ Inducing the Sweep Feature
◊ Inducing the Blend Feature
◊ Inducing the Rib Feature
◊ Modifying Features
◊ Deleting a Feature

In Pro/ENGINEER Wildfire 4.0 the process of a model creation passes through the **Sketch** mode, the **Part** mode, the **Assembly** mode, and the **Drawing** mode. The **Sketch** mode is discussed in Chapter 3, "Exploring the Pro/ENGINEER Wildfire 4.0 Sketch Mode." In this chapter, we discuss the **Part** mode. The **Part** mode in Pro/ENGINEER Wildfire 4.0 is used to build three-dimensional (3D) solid features such as holes, cuts, and rounds, and has properties such as volume, mass, and surface area.

A feature in Pro/ENGINEER Wildfire 4.0 is considered a basic entity. When multiple features are combined together (by keeping some design intent in mind), a component is created. Further, when multiple components are combined, a 3D model is created. Therefore, to create a component of the model, you first have to build a base feature. The base feature is the first solid feature of a model. However, to create features in Pro/ENGINEER Wildfire 4.0, you need to start the **Part** mode of Pro/ENGINEER Wildfire 4.0.

We start this chapter by understanding the process of starting the **Part** mode. Next, we discuss the processes to create features such as cuts, holes, rounds, and chamfers. Finally, after creation, we learn to modify and delete features with regard to other features. Now, let's start studying this chapter. First, we learn the process to start the **Part** mode.

4.1 STARTING THE PART MODE

The **Part** mode is used to build solid features of a model. To start the **Part** mode, you first have to start Pro/ENGINEER Wildfire 4.0. The process to start Pro/ENGINEER Wildfire 4.0 is discussed in Chapter 2, "Exploring the User Interface." Now, let's follow the steps given here to start the **Part** mode:

1. *Select* the **File menu** on the **Menu Bar** in the Pro/ENGINEER Wildfire 4.0 main window, as shown in **Figure 4.1**. The **File menu** options appear (Figure 4.1).
2. Now, *click* the **New** option, as shown in Figure 4.1:

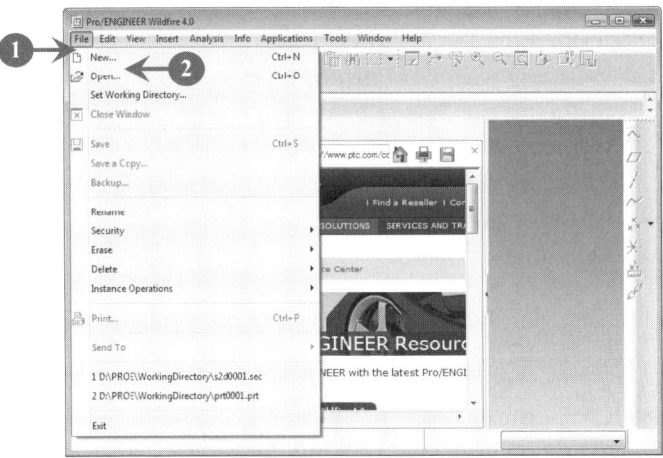

FIGURE 4.1

The **New dialog box** appears (**Figure 4.2**). By default, the **Part** radio button is selected.

3. Assign a name for the **Part** you have to create. The part either accepts the default name or you can type any other name in the text box beside the **Name** label. In our case, we have accepted the default name, prt0001, as shown in Figure 4.2.

4.   Next, *click* the **OK** button, as shown in Figure 4.2:

FIGURE 4.2

The Pro/ENGINEER Wildfire 4.0 window with the **Part** mode appears, as shown in
Figure 4.3:

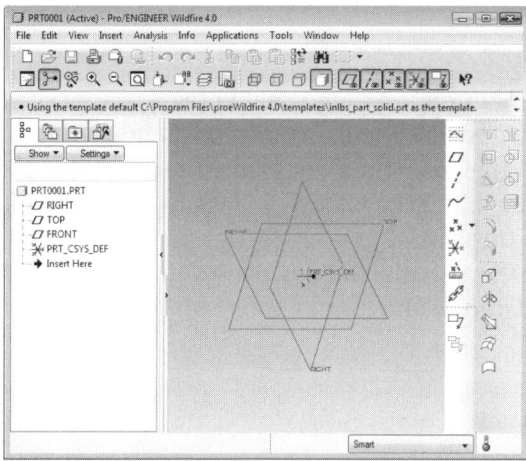

FIGURE 4.3

Now, the **Part** mode is ready to use. Figure 4.3 is the initial window, which appears in
the **Part** mode. This initial window shows three datum planes that are created by default
when you start the **Part** mode. **Datum** planes are important because they help in creating
a base feature. Therefore, before proceeding ahead in this chapter with feature creation,
we should understand datums.

4.2 UNDERSTANDING DATUMS

Datums are imaginary features known as features having no mass, volume, or surface
area. It is important to understand **Datums** because they work as a reference for

sketching a feature, help in orienting components of a model, assemble components, and so on. Since complex models contain many components encapsulating multiple features, **Datums** help in creating complex models. However, every model, whether simple or complex, consists of a base feature on which other features are added. It is not easy to add all the features on a single plane. Therefore, to add other features, you may either select other available planes or create additional planes. By default, the following datum planes are created when you enter the **Part** mode:

- **Right:** Represents the right view of a feature or model.
- **Top:** Represents the top view of a feature or model.
- **Front:** Represents the front view of a feature or model.

These default datum planes are the first feature created in the **Part** mode and all are perpendicular to each other. These planes are used to draw a 2D sketch of a feature and to convert that sketch into a 3D feature through extrusion.

Note: We will study about extrusion in detail later in this chapter.

Although these default datum planes help create a 3D feature, they are not inter-related to each other. These datum planes are independent of each other; that is, if you select the **Front** plane, then only the front plane is selected, not all planes. However, you can select a datum plane either by *selecting* its name in the **Model Tree** or by *clicking* the plane in the **Drawing Area**.

Apart from using the default **Datum** planes to group multiple features, you can also create your own datum planes as per your requirements. Next, let's learn to create datum planes.

Creating Datum Planes

Datum planes can be created by:

- Selecting the **Plane** option in the **Model Datum** option available in the **Insert menu.**
- Using the **Plane tool** available on the **Datum toolbar.**

Note: The **Datum toolbar** is explained in Chapter 2, "Exploring the User Interface."

In this chapter, we create additional datum planes using the **Datum toolbar.** Let's follow the steps given here to create a datum plane:

1. *Start* the **Part** mode. The process to start the **Part** mode is discussed in the previous section.
2. Now, *click* the **Plane** ( ) tool (Figure 4.3) on the **Datum toolbar** in the **Part** mode window. The **DATUM PLANE dialog box** appears, as shown in **Figure 4.4**:

FIGURE 4.4

The **DATUM PLANE dialog box** contains the following tabs to create a datum plane:

- **Placement Tab:** When the **DATUM PLANE dialog box** appears, the **Placement** tab is already selected. Following are the options that the **Placement** tab contains:
  - **Reference Area:** Reference area (Figure 4.4) displays the already existing datum plane, which is used to create a new datum plane. When you select a datum plane in the **Drawing Area**, its configuration appears in the **References** area. Various constraints are automatically applied on the new datum plane depending upon the constraint of the reference plane. These constraints appear on the right of the reference plane. Following are various constraints that are available with the reference planes:
    - ◊ **Offset:** This is a default constraint of a datum plane. It is used to create a datum plane at an offset from the selected reference.
    - ◊ **Through:** This constraint helps create the new datum plane through the selected reference.
    - ◊ **Parallel:** This constraint helps create the new datum plane parallel to the selected reference.

◇ **Normal:** This constraint helps create the new datum plane normal to the selected reference.

◇ **Tangent:** This constraint helps create the new datum plane tangent to the cylindrical features.

• **Offset area:** This area is visible only when the constraint of the reference plane is the **Offset** constraint.

■ **Display Tab:** The options available in the **Display** tab help flip the datum plane and set the size of the datum plane. The **Display** tab is shown in **Figure 4.5**:

FIGURE 4.5

Following are the options available on the **Display** tab:

• **Flip:** This button helps change the normal direction of the datum plane that is to be created.

• **Adjust Outline:** This option helps adjust the perimeter of the datum plane being created. This option is available when the check box (**Figure 4.5**) beside the **Adjust Outline** label is selected. When it is selected, then the text boxes beside the **Width** and **Height** labels are available to specify the width and height of the datum plane, respectively. Apart from this, the combo-box named **Size** also enables adjustments.

• **Lock Aspect Ratio:** This option is available only when the check box beside the **Adjust Outline** label is selected. This option helps manage the proportion between the width and the height of the datum plane outline display.

■ **Properties Tab:** This tab shows the name of the datum plane being created. By default, **DTM1** is assigned to the first datum plane created. When more than one datum plane is created, they are successively numbered. This tab also allows you to assign a new name to the datum plane being created. The **Properties** tab is shown in **Figure 4.6**:

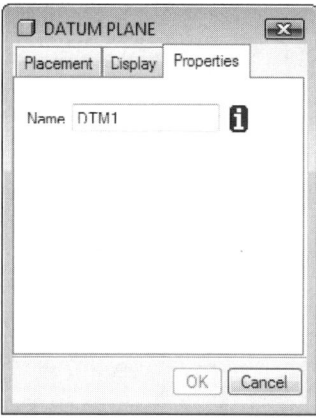

FIGURE 4.6

This section is about the **DATUM PLANE dialog box** and the constraints available to create a datum plane. As discussed, we have various constraints that help us to create a datum plane. However, in this chapter, we use the **Offset** constraint to create a new datum. Therefore, let's continue creating a datum plane.

3. *Select* an existing datum plane (Figure 4.3). The selected datum plane appears in the **References** area in the **DATUM PLANE dialog box**, as shown in **Figure 4.7**:

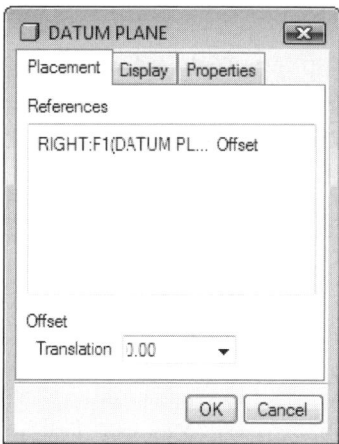

FIGURE 4.7

In our case, we have selected the **Right** datum plane. In Figure 4.7, you can see that the **Offset** area is enabled because the constraint of the reference plane that we have selected is **Offset**.

Note: In case the constraint of the referenced plane is not **Offset**, click the constraint appearing on the right of the selected datum plane in the **DATUM PLANE dialog box**. A drop-down list of constraints appears. Now, select the **Offset** constraint from the list.

Now, you have to set the offset distance of the datum plane being created.

4. To set the offset distance, *click* and *drag* the offset handle to move the datum plane to the required distance, as shown in **Figure 4.8**:

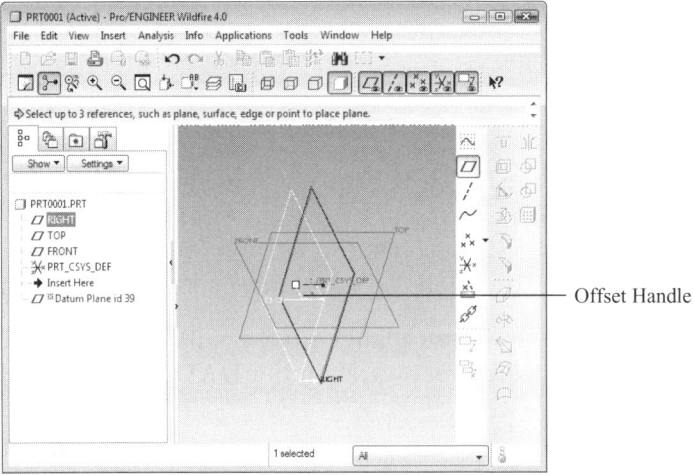

FIGURE 4.8

Note: You can also set the offset distance by specifying the offset value in the drop-down list beside the **Translation** label (Figure 4.7) or select any value from the recently used values.

In Figure 4.8, the white-colored datum plane is the new datum plane created. The value of the offset distance is automatically set in the drop-down list beside the **Translation** label in the **DATUM PLANE dialog box**, as shown in **Figure 4.9**.

5. Now, *click* the **OK** button (**Figure 4.9**) to complete the datum plane creation with the default name:

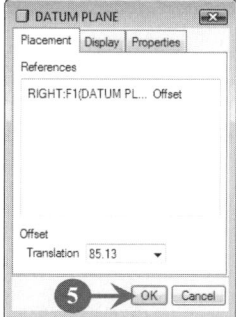

FIGURE 4.9

Note: However, if you want to assign a new name to the datum plane, then click the **Properties** tab in the **DATUM PLANE dialog box**. Specify the new name to the datum plane and *click* the **OK** button. In our case, we have accepted the default name.

As soon as you *click* the **OK** button, the new datum plane appears in the **Drawing Area** and also in the **Model Tree**, as shown in **Figure 4.10**:

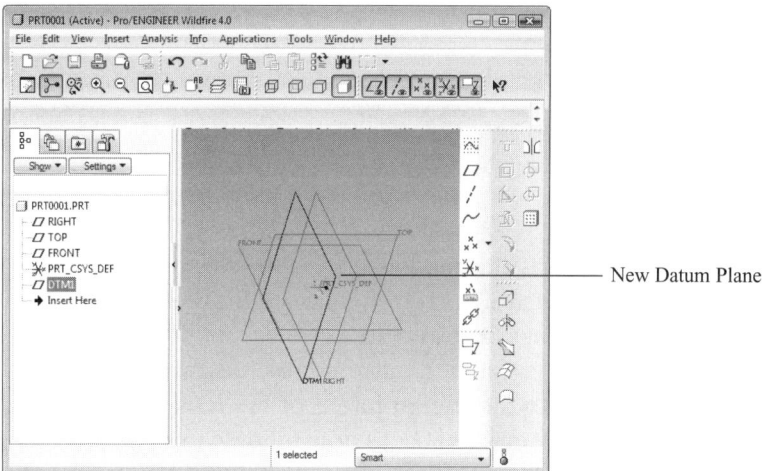

FIGURE 4.10

In Figure 4.10, you can see the new datum plane with the name **DTM1** appearing in the **Model Tree** as well as the **Drawing Area**. This section is about creating a datum plane in Pro/ENGINEER Wildfire 4.0. Next, we start creating features.

4.3 CREATING AN EXTRUDED FEATURE

Extrusion is a process of creating a 3D model of a 2D sketch. To create an extruded feature, the **Extrude** (⬚) button available on the **Base Features toolbar** in the **Part** mode window can be used. The **Base Features toolbar** is shown in **Figure 4.11**:

FIGURE 4.11

An extruded feature can also be created using the **Extrude** option in the **Insert menu**. The **Extrude** button is generally used for two purposes:

- Creating a Base Feature.
- Creating a Protrusion.

Let's now discuss these purposes one by one.

Creating a Base Feature

A **base feature** is the basic entity of a model because the shape of a model depends on its base. For example, if the base feature is rectangular, then a rectangular model is created; and if the base feature is circular, it results in a circular model. All other features such as protrusions, cuts, and holes are added onto the base feature to complete a model. A base feature varies according to the design intent of the model being designed. In our case, a rectangular slot is being created. Let's now follow the steps given here to create a base feature:

1. *Start* the **Part** mode.
2. Now, *click* the **Extrude** button in the **Base Features** toolbar (Figure 4.11). The **Extrude** dashboard appears, as shown in **Figure 4.12**:

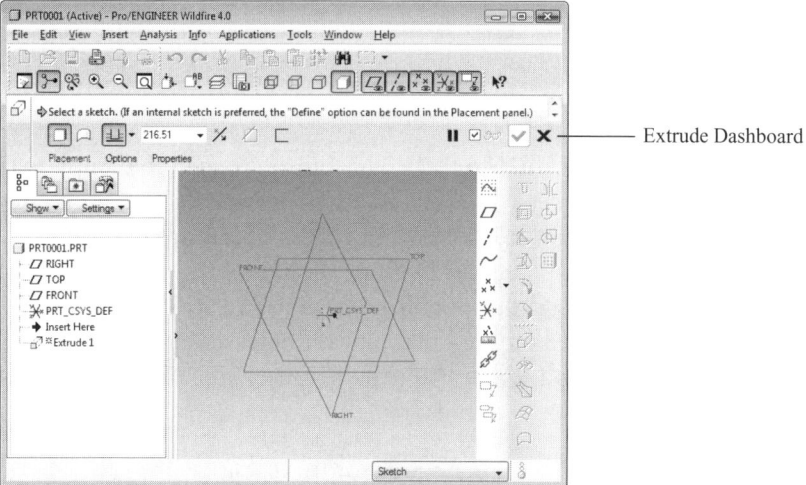

Extrude Dashboard

FIGURE 4.12

Let's have a brief look at the buttons and tabs available in the **Extrude** dashboard before moving ahead with the steps in this section. The **Extrude** dashboard contains the following buttons and tabs:

- **Extrude as Solid Button ( ):** Used to create a solid feature.
- **Extrude as Surface Button ( ):** Used to create the surface of a model.
- **Depth Options ( ):** Used to constrain the depth of a feature. When you *click* the down arrow in the **Extrude** dashboard, various depth options project out. You can use any of the depth options to constrain the depth of the feature being created.
- **Depth Box (149.52):** Specifies the depth of the feature being created. You can either select the available depth by *clicking* the down arrow beside (149.52) in the **Extrude** dashboard or specify any other required depth.
- **Flip Button ( ):** Flips the direction of the feature being created with respect to the **Sketching** plane. This button is enabled only when the feature creation is completed.
- **Remove Material Button ( ):** Removes the material from the feature. This button is enabled only when the feature creation is completed.
- **Thicken Sketch Button ( ):** Assigns specific thickness to the sketch outline.
- **Pause Button ( ):** Used to pause the process of current feature creation.

- **Resume Button (▶):** Resumes the process of feature creation. This button is enabled only when the **Pause** button is clicked.
- **Preview Button (☑👓):** This button is used to preview the feature being created.
- **Build Feature Button (☑):** This button is used to complete the feature creation.
- **Close Button (✖):** Aborts the process of feature creation.
- **Placement Tab (Placement):** Defines the sketching plane.
- **Options Tab (Options):** Specifies whether an extrusion is to be created on one side of the sketching plane or on both sides.
- **Properties Tab (Properties):** Edits the feature name. This tab can also be used to know the feature details.

This section is briefly about the **Extrude** dashboard. Let's now move ahead with the process of creating the base feature.

3. *Select* the **Placement** tab (**Figure 4.13**) to define the sketching plane. A slide-down panel, as shown in Figure 4.13 appears:

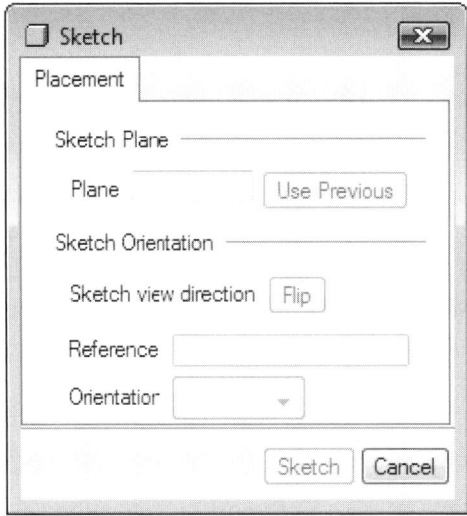

FIGURE 4.14

4. *Click* the **Define** button in the slide-down panel (Figure 4.13). The **Sketch Dialog Box**, as shown in **Figure 4.14**, appears.
5. *Select* a datum plane either in the **Model Tree** or in the **Drawing Area** (Figure 4.13). Once you have selected the datum plane, the selected datum plane, its reference plane, and orientation are automatically specified in the **Sketch dialog box**, as shown in **Figure 4.15**.

In our case, we have selected the **Front** datum plane. In Figure 4.15, you see that the selected datum plane is specified in the text box beside the **Plane** label, its reference is specified in the text box beside the **Reference** label, and the orientation appears beside the **Orientation** label.

6. Next, *click* the **Sketch** button (Figure 4.15) to confirm the selection, as shown in Figure 4.15:

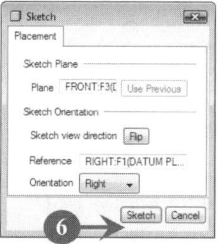

FIGURE 4.15

Now, the **Sketcher tools toolbar** appears, as shown in **Figure 4.16**:

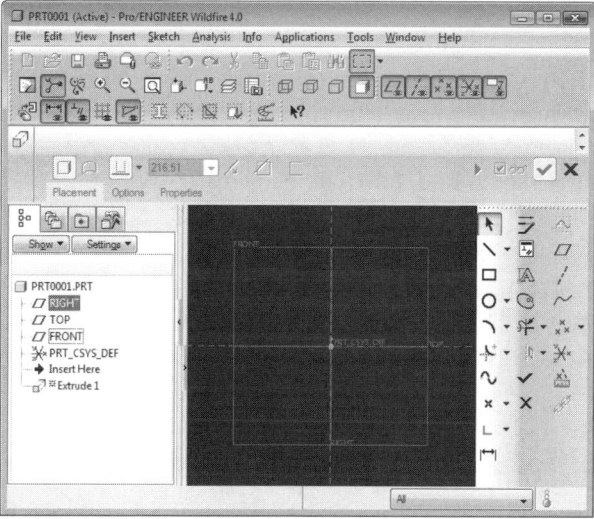

FIGURE 4.16

Now, you can create a sketch of the base feature. In our case, we created a rectangle. After creating a rectangle, apply dimensions on it.

Note: The process to sketch a rectangle and applying dimensions on it is depicted in Chapter 3, "Exploring Pro/ENGINEER Wildfire 4.0 Sketch Mode."

Once the dimensions are applied on the rectangle, it appears as shown in **Figure 4.17**.

7.   Now, *click* the **Done** (✔) button in the **Sketcher tools toolbar** (Figure 4.17):

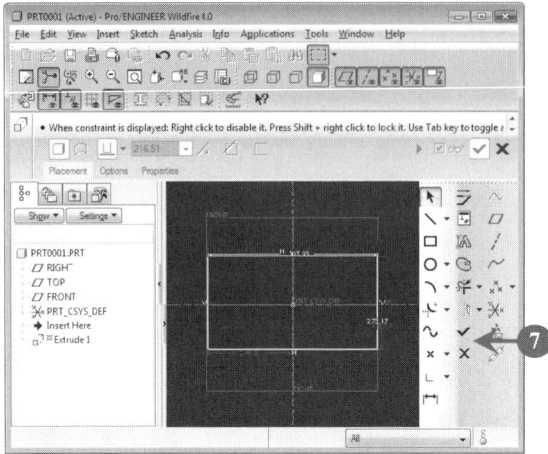

FIGURE 4.17

The sketch will appear in yellow on your screen (**Figure 4.18**). Now, you can view the sketch in 3D mode using various orientations, such as **Left** or **Right**.

8.   To view the sketch in 3D format using different orientations, *click* the **Named View List** (⬚) button in the **View toolbar**, as shown in Figure 4.18:

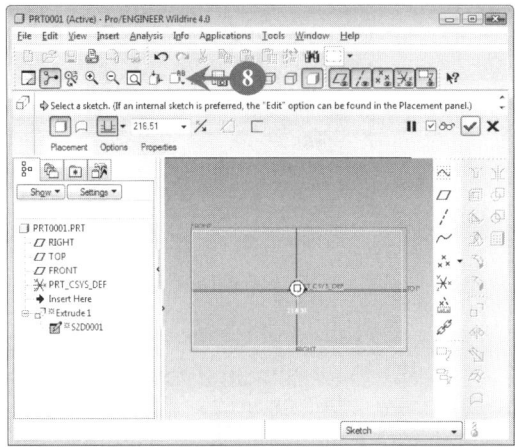

FIGURE 4.18

A list of various orientations appears (**Figure 4.19**).

9. *Select* any of the orientations to view the sketch in 3D format. In our case, we have selected the **Default Orientation**, as shown in Figure 4.19:

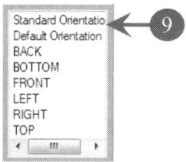

FIGURE 4.19

Once the **Default Orientation** has been selected, the rectangular slot created appears, as shown in **Figure 4.20**:

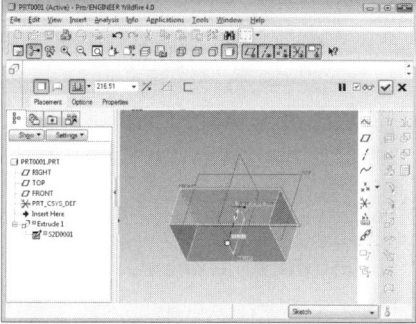

FIGURE 4.20

10. Now, *specify* its depth. To specify its depth, either select a previously used value from the **Depth box** or enter a new value in the **Depth box**. In our case, we entered 100.00 as the new value.
11. *Press* the **ENTER** key. The 3D view of the rectangular slot appears (**Figure 4.21**).
12. *Click* the **Preview** (⬚⬚) button in the **Extrude** dashboard to preview the rectangular slot being created, as shown in Figure 4.21:

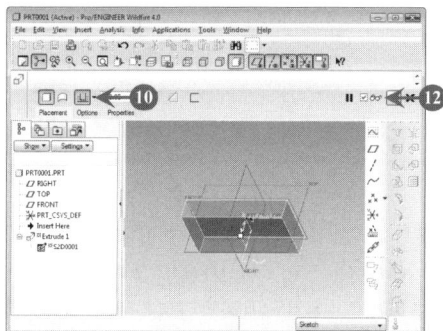

FIGURE 4.21

The preview of the rectangular slot appears (**Figure 4.22**).

13. *Click* the **Build Feature** (✓) button in the **Extrude** dashboard to complete the base feature creation, as shown in Figure 4.22:

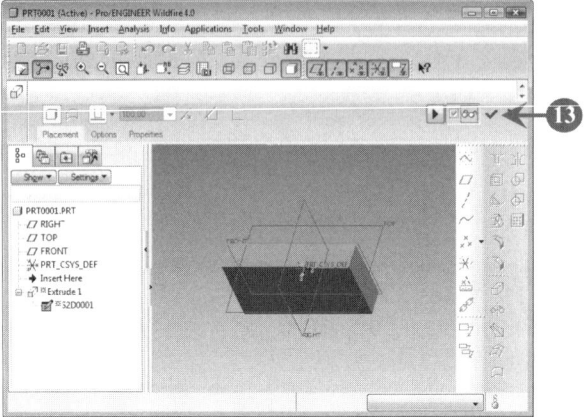

FIGURE 4.22

As soon as you *click* the **Build Feature** button, the rectangular slot appears, as shown in **Figure 4.23**:

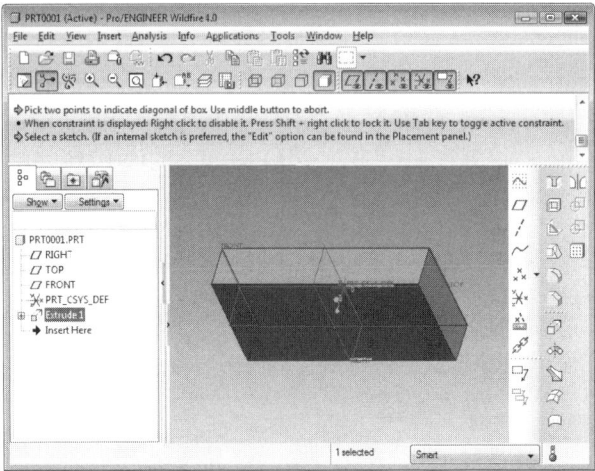

FIGURE 4.23

14. Now, *click* anywhere on the **Drawing Area** (Figure 4.23). The final view of the base feature appears, as shown in **Figure 4.24**:

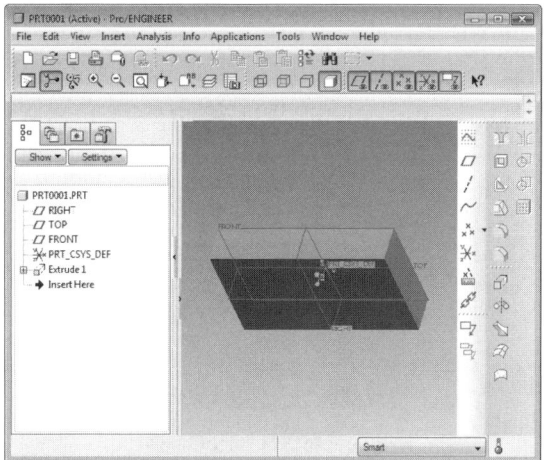

FIGURE 4.24

15. *Save* the feature. The process to save a feature is depicted in Chapter 2,
    "Exploring the User Interface."

Note: A feature created in the Part mode is saved as a .PRT file.

In this way, a base feature is created. Next, we learn the process to create
protrusions.

Creating a Protrusion

Protrusion can be defined as the material added to the base feature. The shape of the
protrusion depends on the design of the model being created. To create a protrusion, a
base feature must exist. In our case, the rectangular slot created is the base feature in the
section, "Creating a Base Feature." Let's now follow the steps to create a protrusion:

1. *Start* the **Part** mode and open the .PRT file containing the rectangular slot
   to create a protrusion on it. In our case, the file name is PRT0001.PRT.
2. To open the PRT0001.PRT, *click* the **File > Open**. The **File Open dialog
   box** appears.
3. Now, *select* the PRT0001.PRT file in the **File Open dialog box.**
4. *Click* the **OK** button. The PRT0001.PRT file opens in the **Part** mode, as
   shown in **Figure 4.25**.

In Figure 4.25, you can view the rectangular slot created as a base feature in the
section, "Creating a Base Feature."

5. *Click* the **Extrude** button in the **Base Features toolbar**, as shown in Figure 4.25:

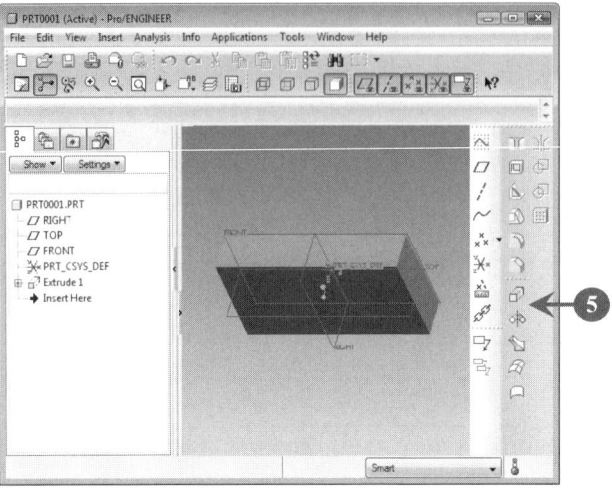

FIGURE 4.25

The **Extrude** dashboard appears (Figure 4.12).

6. Next, *select* the **Placement** tab (Figure 4.12). A slide-down panel that lets you define a sketching plane appears (Figure 4.13).
7. *Click* the **Define** button on the slide-down panel (Figure 4.13). The **Sketch dialog box** appears (Figure 4.14).
8. *Select* a datum plane either in the **Model Tree** or in the **Drawing Area**. In our case, we have selected the **Front** datum plane. Once you have selected the datum plane, the selected datum plane and its reference are automatically specified in the **Sketch dialog box**, as shown in **Figure 4.26**.
9. *Click* the **Sketch** button (Figure 4.26) to confirm selection:

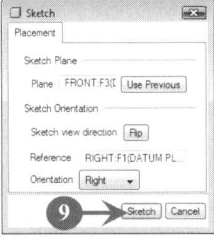

FIGURE 4.26

Now, the **Sketcher tools toolbar** is enabled, as shown in **Figure 4.27**:

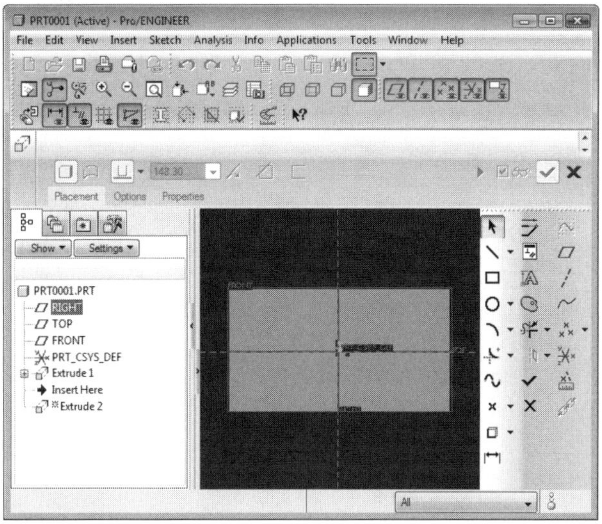

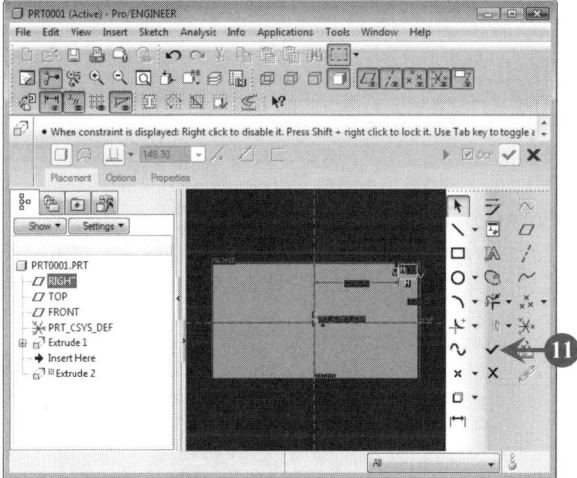

FIGURE 4.27

10. *Draw* a small rectangle at the upper right corner of the rectangular slot (Figure 4.28).
11. *Click* the **Done** (✓) button on the **Sketcher tools toolbar**, as shown in **Figure 4.28**:

FIGURE 4.28

The preview of the newly created sketch appears in yellow (**Figure 4.29**).

12. *Click* the down arrow beside the **Named View List** (🖼) button in the **View toolbar** to view the feature in 3D mode, as shown in Figure 4.29:

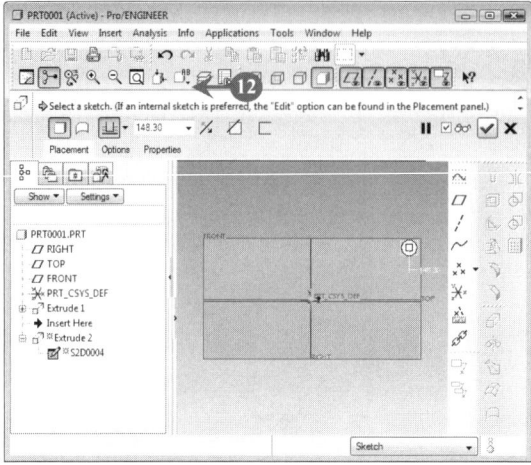

FIGURE 4.29

A list of various orientations appears (Figure 4.19).

13. *Select* the **Default Orientation** (Figure 4.19). The rectangular slot with the protrusion appears, as shown in **Figure 4.30**.

Modify the depth of the protrusion, if required. In our case, we are continuing with the default values.

14. *Click* the **Build Feature** button in the **Extrude** dashboard, as shown in Figure 4.30:

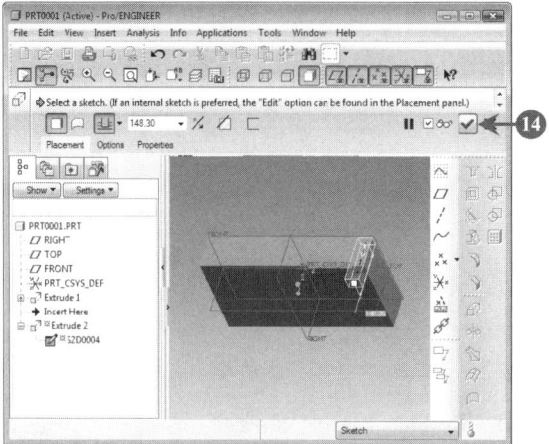

FIGURE 4.30

The rectangular slot with a protrusion appears, as shown in **Figure 4.31**:

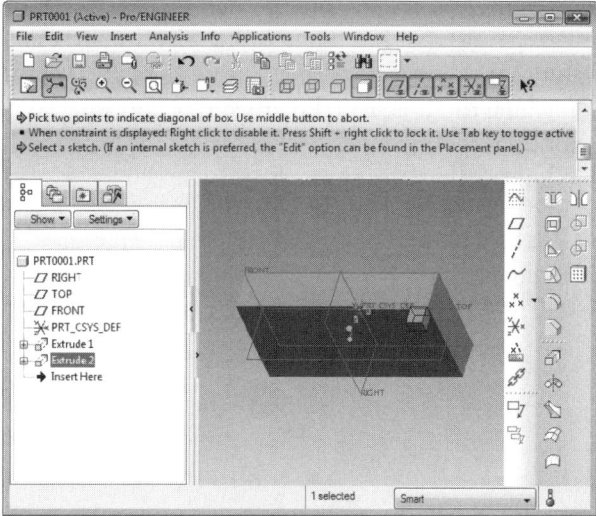

FIGURE 4.31

15. Now, *click* anywhere in the **Drawing Area** (Figure 4.31). The final view of the base feature along with the protrusion appears, as shown in **Figure 4.32**:

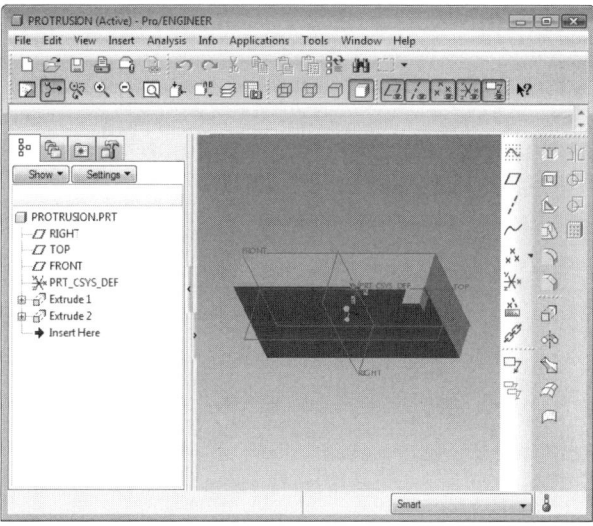

FIGURE 4.32

In this way, a protrusion is created. Let's move ahead by adding another feature in the base feature.

4.4 INDUCING THE HOLE FEATURE

In Pro/ENGINEER Wildfire 4.0, simple and industry-standard holes can be created. An industry-standard hole is a hole that has some industry standards to define its diameter. In this section, we see the process to create a simple hole. To create a simple hole, we need a base feature on which the hole is to be created. Therefore, we use the rectangular slot created in the section, "Creating a Base Feature" as the base feature on which a hole is created. To create a hole, follow the steps given:

1. *Start* the **Part** mode and *open* the PRT0001.PRT file.
2. *Click* the **Hole** (▼) button in the **Engineering Features toolbar** in the **right tool chest**. The **Engineering Features toolbar** is shown in **Figure 4.33**:

FIGURE 4.33

The **Hole** dashboard appears on top of the drawing area, as shown in **Figure 4.34**:

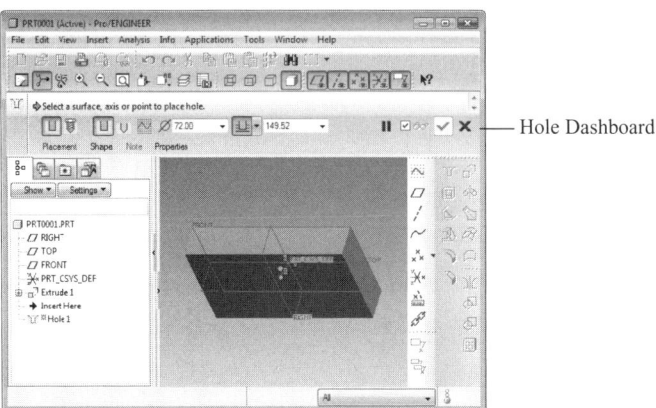

Hole Dashboard

FIGURE 4.34

Note: As soon as the **Hole** dashboard appears, the **Engineering Features toolbar** is disabled.

3. *Select* an area on the rectangular slot where the hole is to be created. The preview of the hole is displayed, as shown in **Figure 4.35**.

In our case, we selected the middle area of the rectangular slot. The hole to be created is previewed in yellow color (Figure 4.35).

4. *Select* the **Placement** tab on the **Hole** dashboard, as shown in Figure 4.35:

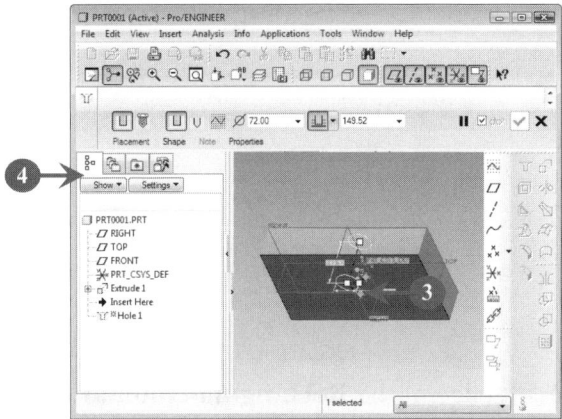

FIGURE 4.35

The **Placement** tab is selected to choose the sketching plane and its offset references. The **Placement** slide-down panel appears, as shown in **Figure 4.36**.
    In Figure 4.36, the plane selected on the rectangular slot appears in the text box in the **Placement** option.

5. *Click* inside the text area in the **Offset References** option to specify the offset references, as shown in Figure 4.36:

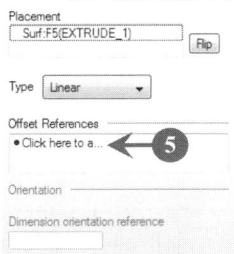

FIGURE 4.36

In the text area, the message "Select 2 items" is flagged. This message prompts you to select two items to specify the offset references, as shown in **Figure 4.37**:

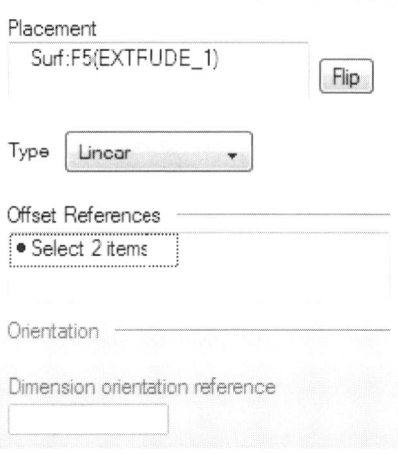

FIGURE 4.37

6. *Select* the left edge of the **Rectangular toolbar**. The configuration of the selected edge appears in the **Offset References** edit box. Apart from that, you are also prompted to select the remaining item as an offset reference, as shown in **Figure 4.38**:

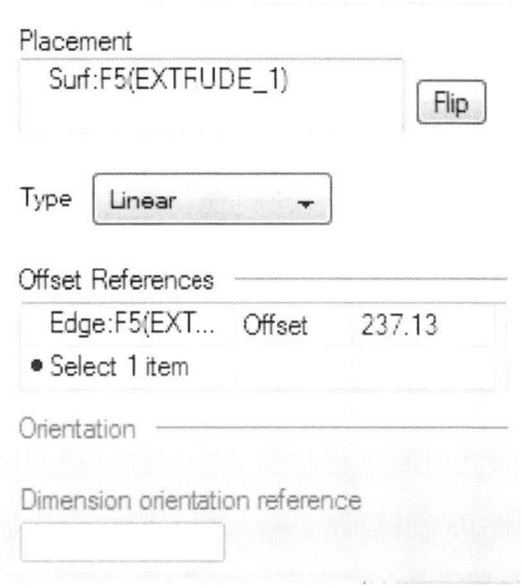

FIGURE 4.38

7. *Press* the **CTRL** key and *select* the top edge of the rectangular slot. Both selected edges appear in the edit box in the **Offset Reference** option. Apart from that, both edges are highlighted in red (**Figure 4.39**).

8. *Select* the **Placement** tab once again (Figure 4.39) to close the **Placement** slide-down panel.

9. *Click* the down arrow in the **Depth** ⬜▾ box to *select* the **Predefined Depth** option, as shown in Figure 4.39:

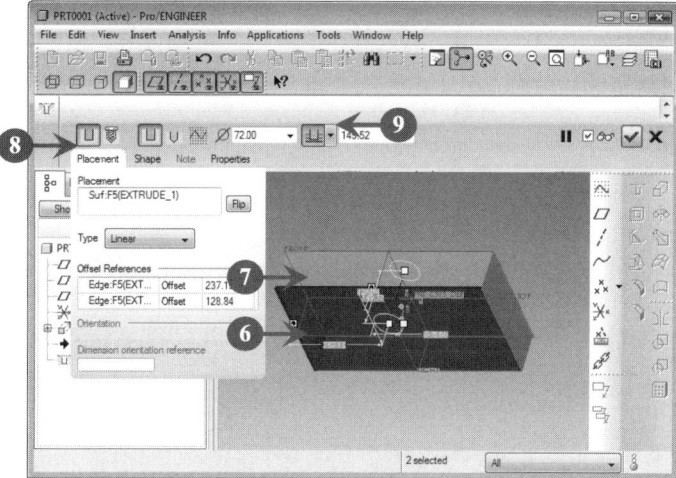

FIGURE 4.39

A list of available depth options appears (**Figure 4.40**).

10. *Select* the **Drill to intersect with all surfaces** (⬛) option from the list, as shown in Figure 4.40:

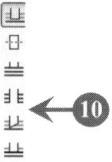

FIGURE 4.40

This option creates a hole by drilling the surface of the rectangular slot.

11. *Click* the **Build Feature** (✅) button (Figure 4.39) to complete the hole creation. A hole, as shown in **Figure 4.41**, is created:

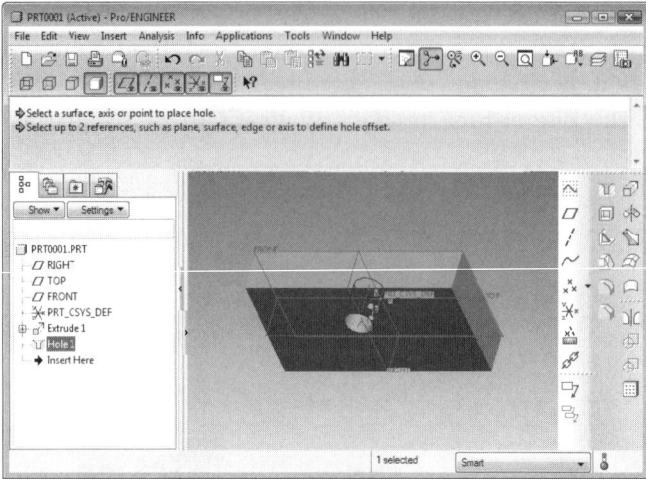

FIGURE 4.41

12. *Click* anywhere in the **Drawing Area** (Figure 4.41). The final view of the base feature along with the hole feature appears, as shown in **Figure 4.42**:

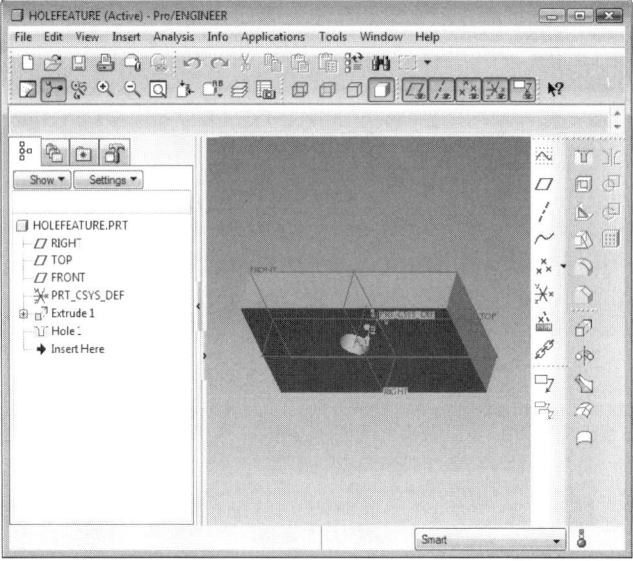

FIGURE 4.42

13. *Save* the file. In our case, we have saved the file with the name HOLEFEATURE.prt.

This is how a hole is created on a feature in Pro/ENGINEER Wildfire 4.0. Let's now move further and depict the process to create a feature called **cut**.

4.5 CREATING A CUT FEATURE

A **cut** in Pro/ENGINEER Wildfire 4.0 can be rectangular, triangular, or semicircular. The shape of a **cut** depends on the sketch created on a feature. A **cut** in Pro/ENGINEER Wildfire 4.0 is created using the **Extrude** dashboard. However, to create a **cut** a 3D feature is required. In our case, we use the HOLEFEATURE.prt file created in the preceding section as the base feature. Now, let's follow the steps given here to create a **cut** feature:

1. *Start* the **Part** mode and *open* the HOLEFEATURE.prt file (Figure 4.42).
2. *Click* the **Extrude** (⬚) button in the **Base Features toolbar** (Figure 4.11) in the **right tool chest**. The **Extrude** dashboard appears, as shown in **Figure 4.43**.

Note: Figure 4.43 also displays the HOLEFEATURE.prt file.

3. *Click* the **Remove Material** (⬚) button on the **Extrude** toolbar (Figure 4.43).
4. *Select* the **Placement** tab to define the sketching plane, as shown in Figure 4.43. As soon as you select the **Placement** tab, a slide-down panel appears (Figure 4.13).
5. Next, *click* the **Define** button in the slide-down panel (Figure 4.13). The **Sketch dialog box** appears (Figure 4.14).
6. *Select* the datum plane or *click* on the feature's surface to create a cut. In our case, we have *clicked* the right side of the hole to specify the surface, as shown in **Figure 4.43**:

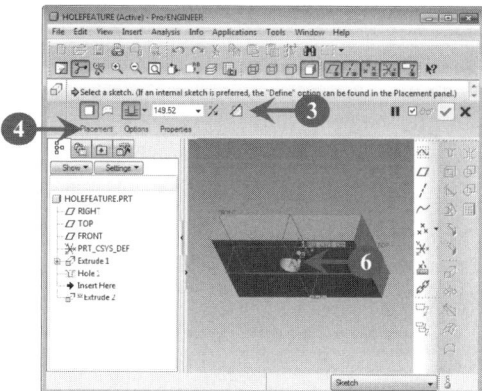

FIGURE 4.43

The selected surface, its reference plane, and orientation automatically appear in the **Sketch dialog box** (**Figure 4.44**).

7. *Click* the **Sketch** button to confirm the selection, as shown in Figure 4.44:

FIGURE 4.44

The **Sketcher tools toolbar** appears (**Figure 4.45**).

8. *Select* the **Rectangle** button on the **Sketcher tools toolbar** (Figure 4.45).
9. *Draw* another rectangle on the right side of the hole (Figure 4.45).
10. *Click* the **Done** (☑) button in the **Sketcher tools toolbar**, as shown in Figure 4.45:

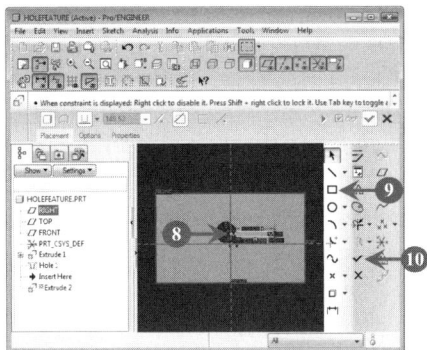

FIGURE 4.45

The sketch appears in yellow (**Figure 4.46**).

11. *Click* the **Preview** (☑ẟ) button in the **Sketcher tools toolbar**, as shown in Figure 4.46:

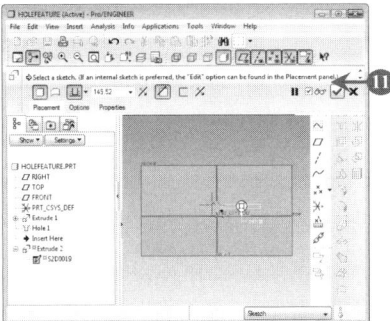

FIGURE 4.46

The cut feature is previewed (**Figure 4.47**).

12. *Click* the **Build Feature** (☑) button in the **Extrude** dashboard (Figure 4.47) to complete the cut creation:

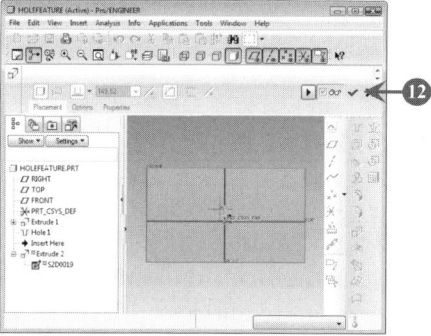

FIGURE 4.47

The final 3D cut feature is shown in **Figure 4.48**:

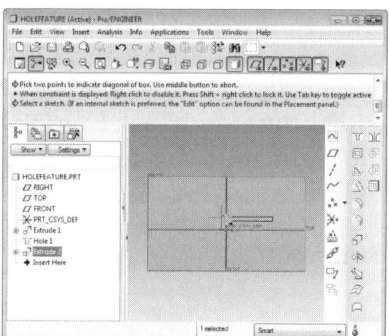

FIGURE 4.48

The edges of the cut we have created are highlighted in red.

13. *Click* anywhere in the **Drawing Area** (Figure 4.48). The final view of the base feature along with the cut feature appears, as shown in **Figure 4.49**:

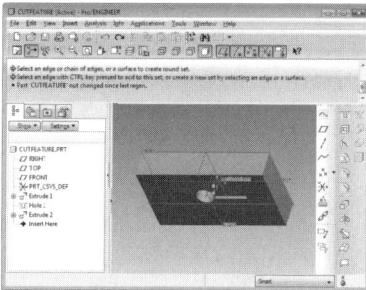

FIGURE 4.49

14. *Save* the file. In our case, we have saved the file with the name CUTFEATURE.prt.

In this way, a cut feature is created on any other feature to get the desired 3D component. Now, we study the process to add a round feature on a preexisting feature.

4.6 INDUCING THE ROUND FEATURE

The round feature is used to create a round transition between two adjacent edges. In Pro/ENGINEER Wildfire 4.0 the **Round** dashboard is used to add a round feature to an existing feature. You can invoke the **Round** dashboard in two ways: first, by selecting the **Round** option in the **Insert** menu; and second, by clicking the **Round** button in the **Engineering Features toolbar**. In this section, we have used the second way to invoke the **Round** dashboard. Apart from that, we need a feature on which the round feature is to be added. Therefore, we use the CUTFEATURE.prt file created in the preceding section to add a round feature on it. Let's follow the steps given here to add a round feature on a preexisting feature:

1. *Start* the **Part** mode and open the CUTFEATURE.prt file (Figure 4.49).
2. *Click* the **Round** (🖿) button on the **Engineering Features toolbar** (Figure 4.33) in the **right tool chest**. The **Round** dashboard appears, as shown in **Figure 4.50**.

Apart from the **Round** dashboard, the part file CUTFEATURE.prt is also displayed in Figure 4.50.

3. *Select* the **Sets** tab in the **Round** dashboard, as shown in Figure 4.50:

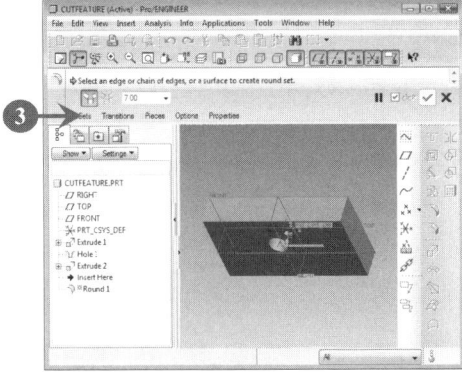

FIGURE 4.50

A slide-down panel appears, as shown in **Figure 4.51**:

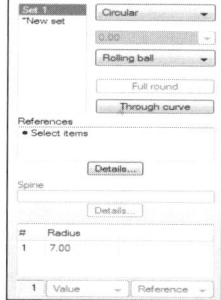

FIGURE 4.51

This slide-down panel allows you to select the edges of a feature on which the round feature has to be applied. The edges you select are grouped in sets and those sets are numbered as **Set1**, **Set2**, and so on. You can create multiple sets by clicking the ***New set** in the slide-down panel to group different edges. By default, **Set1** is created and all the edges you select are grouped in this set. In our case, we group all the edges in the default set. The message, "Select items" in the **References** text box, prompts you to select the edges that need to be rounded.

4. *Select* the first edge that you want to round. In our case, we have selected the edge of the circle in the base feature, as shown in **Figure 4.52**:

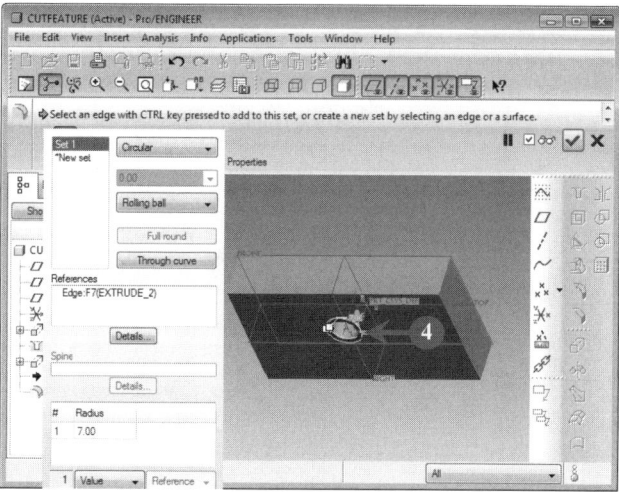

FIGURE 4.52

The selected edge appears in the **References** text box (Figure 4.52).

5. To select more edges, *hold* the **CTRL** key and select edges. In our case, we have selected all the edges appearing in the default orientation of the base feature (**Figure 4.53**).
6. *Press* the middle mouse button and spin the base feature to select the remaining edges, if required. In our case, we have selected all the edges of the rectangular slot.

After selecting all the edges, *set* the radius of the round. By default, some value as the radius of the round appears. In our case 7.00 is the radius of the round, which can be viewed in the edit box of the **Radius** option.

7. *Modify* the radius of the round (Figure 4.53). In our case, we continue with the default value.
8. *Click* the **Build feature** (☑) button to complete adding the round feature, as shown in Figure 4.53:

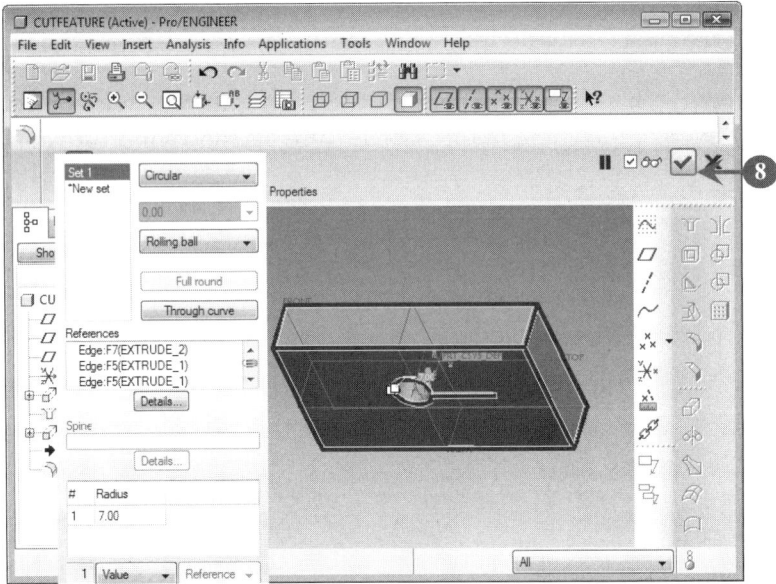

FIGURE 4.53

After adding the round feature, the base feature appears, as shown in **Figure 4.54**:

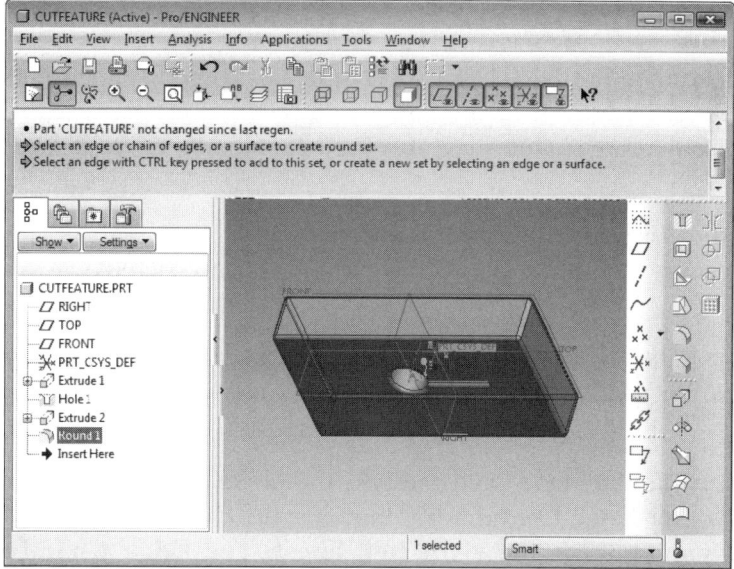

FIGURE 4.54

9. *Click* anywhere in the **Drawing area** (Figure 4.54). The final view of the base feature along with the round feature appears, as shown in **Figure 4.55**:

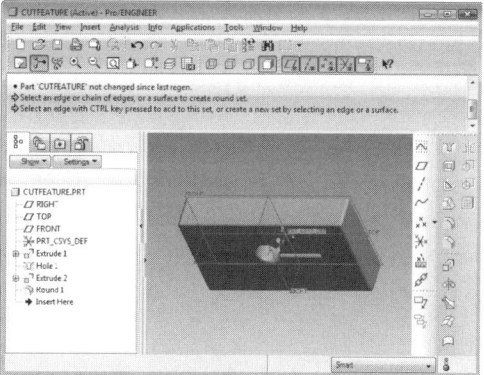

FIGURE 4.55

10. *Finally, save* the file. In our case, we have saved the file with the name ROUNDFEATURE.prt.

This is the process of adding a round feature on a base feature. Next, we learn another feature called chamfer.

4.7 INDUCING THE CHAMFER FEATURE

The chamfer feature in Pro/ENGINEER Wildfire 4.0 is used to create a beveled surface around the selected edges or at the corners of selected edges. In Pro/ENGINEER Wildfire 4.0, two types of chamfer features can be added to a base feature:

- **Edge Chamfer:** Bevels the selected edges
- **Corner Chamfer:** Bevels the corners of the selected edges

In this section, we see the process to add edge chamfer on the edges of the base feature. Let's follow the steps given here to add edge chamfer to a feature:

1. *Start* the **Part** mode and *open* the CUTFEATURE.prt file (Figure 4.49).
2. *Click* the **Edge Chamfer** (◥) button in the **Engineering Features toolbar** (Figure 4.33) in the **right tool chest**. The **Edge Chamfer** dashboard appears, as shown in **Figure 4.56**.

Note: You can also invoke the **Edge Chamfer** dashboard by selecting the **Edge Chamfer** option from the **Chamfer** option in the **Insert menu**.

---

3. *Select* the **Sets** tab in the **Edge Chamfer** dashboard, as shown in Figure 4.56:

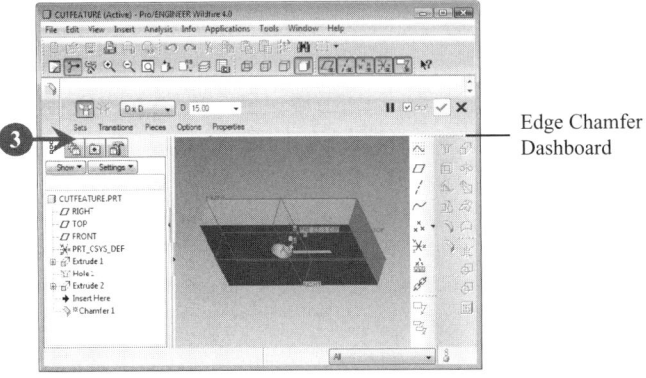

Edge Chamfer Dashboard

FIGURE 4.56

A slide-down panel appears that prompts you to select the edges that need to be chamfered, as shown in **Figure 4.57**:

FIGURE 4.57

4. *Select* the edges you want to bevel in the base feature (Figure 4.56). In our case, we have selected all four top edges in the base feature (**Figure 4.58**).

In Figure 4.58, the number 7.00 appears. This number represents the distance of the edge chamfer from the selected edges. Now, you have two options: either continue with this distance value or modify it. In our case, we are modifying the value.

5. To modify the distance, *double-click* the number, as shown in Figure 4.58:

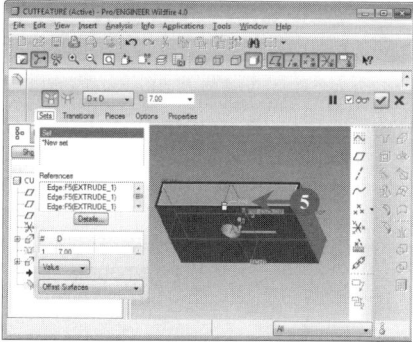

FIGURE 4.58

An edit box appears, as shown in **Figure 4.59**:

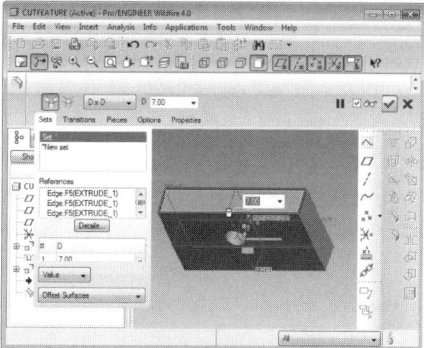

FIGURE 4.59

6. Either *enter* some new value in the edit box or select any previously used value. In our case, we have entered 15.00 as the new value. However, to select any previously used value, *click* the down arrow beside the number and select the value.

7. *Press* the **ENTER** key. The edit box disappears (**Figure 4.60**). The new distance value is highlighted in yellow (Figure 4.60).

8. *Click* the **Build Feature** (☑) button to complete adding **Edge Chamfer** on the base feature, as shown in Figure 4.60:

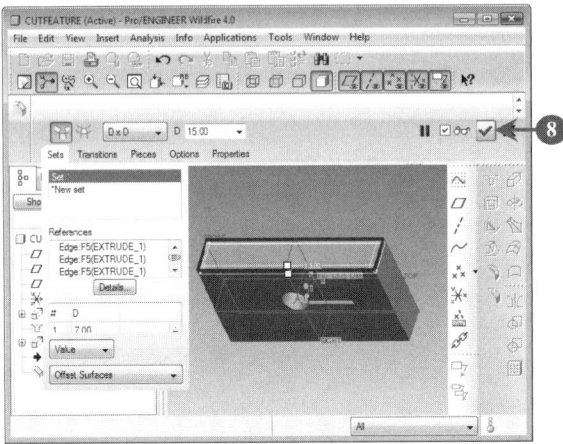

FIGURE 4.60

The slide-down panel disappears and the base feature along with the edge chamfer feature appears, as shown in **Figure 4.61**:

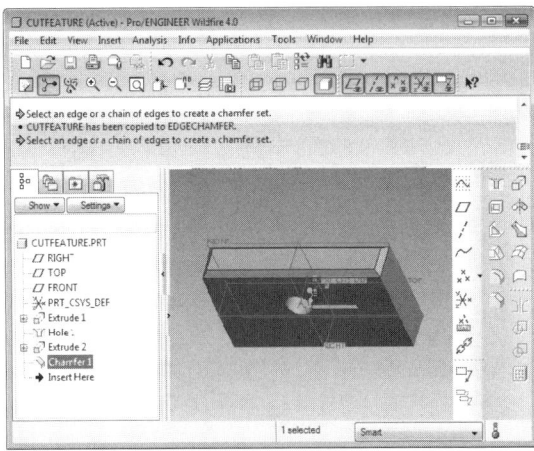

FIGURE 4.61

9. *Click* anywhere in the **Drawing Area**. The final view of the base feature along with the edge chamfer feature appears, as shown in **Figure 4.62**:

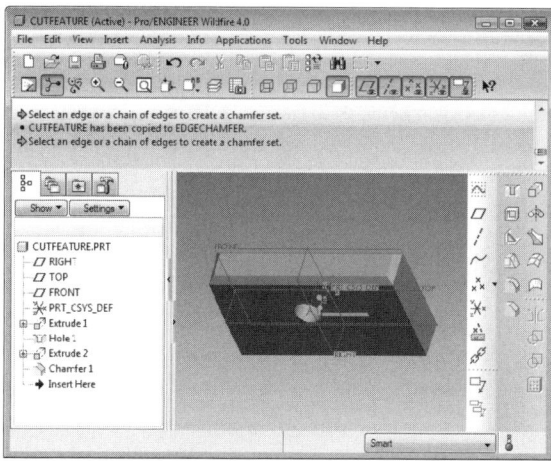

FIGURE 4.62

10. Finally, *save* the file. In our case, we have saved the file with the name EDGEFEATURE.prt.

With this, we have learned the process of adding a chamfer feature on a base feature. Next, we discuss another feature called **Shell**.

4.8 INDUCING A SHELL FEATURE

The shell feature in Pro/ENGINEER Wildfire 4.0 creates a shell by removing a surface from a solid object. A shell created in Pro/ENGINEER Wildfire 4.0 has a specific wall thickness. A shell in Pro/ENGINEER Wildfire 4.0 is created using the **Shell** dashboard. However, to create a shell we need a base feature; therefore, in this section we use the PRT0001.prt file as the base feature. This file contains a rectangular slot. Now, let's follow the steps given here to create a shell:

1. *Start* the **Part** mode and *open* the PRT0001.prt file (Figure 4.24).
2. *Click* the **Shell** (▢) button on the **Engineering Features toolbar** (Figure 4.33) in the **right tool chest.** The **Shell** dashboard appears, as shown in **Figure 4.63**:

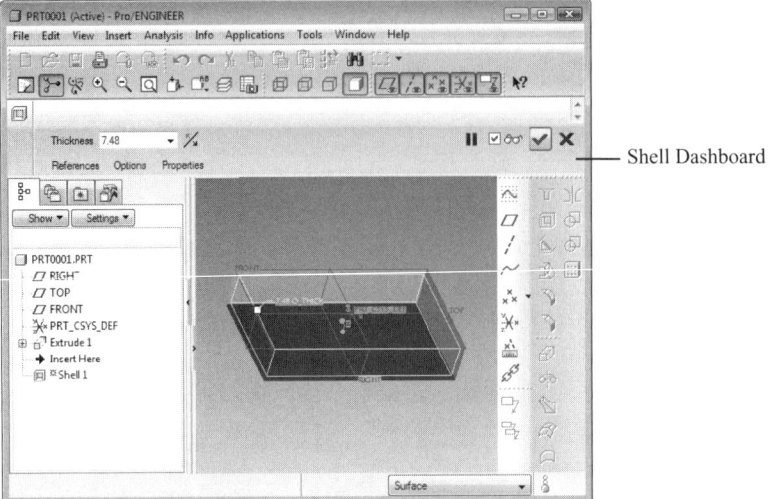

Shell Dashboard

FIGURE 4.63

As soon as the **Shell** dashboard is invoked, all the surfaces of the base feature are highlighted in yellow. Apart from that, the thickness of the walls of the surfaces also appears beside the O_THICK label on the base feature in the **Drawing Area**. In our case, the thickness of the walls of all the surfaces is 7.48. The thickness of the walls can also be viewed in the drop-down edit box beside the **Thickness** label in the **Shell** dashboard (Figure 4.63).

Note: If the thickness of the walls varies then the thickness of the walls appears individually.

3. *Select* the surface that has to be removed from the rectangular slot to create a shell. In our case, we have selected the front surface. As soon as you select the surface, the color of the selected surface changes (**Figure 4.64**).

Now, we have two options: either we can modify the thickness of the walls of the rectangular slot, or continue the shell creation with the default thickness. In our case, we are modifying the thickness value. To modify the thickness, we have two options:

■ *Double-click* the thickness appearing on the base feature in the **Drawing Area** (Figure 4.64).
■ Enter a new thickness value in the drop-down edit box beside the **Thickness** label in the **Shell** dashboard (Figure 4.64).

4. *Double-click* the thickness appearing on the base feature, as shown in Figure 4.64:

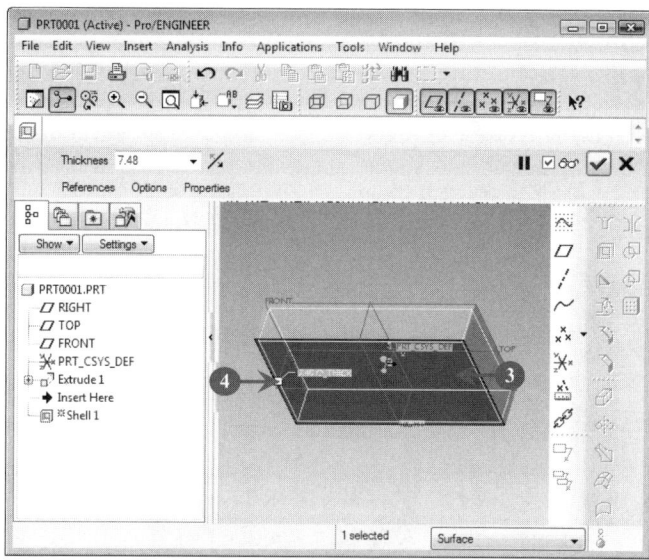

FIGURE 4.64

The thickness of the edit box appears (**Figure 4.65**).

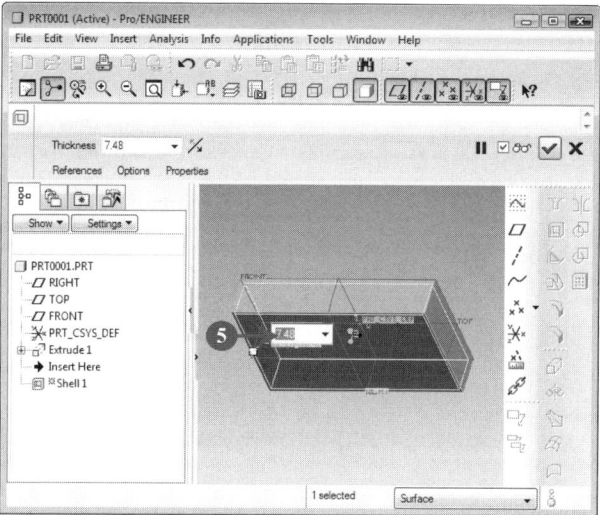

FIGURE 4.65

5. Now, *specify* a new thickness in the edit box (Figure 4.66). In our case, we will enter 25.25 as the new value.

6. *Press* the **ENTER** key. The new thickness is previewed in the **Drawing Area** and also in the drop-down edit box beside the **Thickness** label in the **Shell** dashboard (**Figure 4.66**).

7. *Click* the **Build Feature** (☑) button to complete the shell creation, as shown in Figure 4.66:

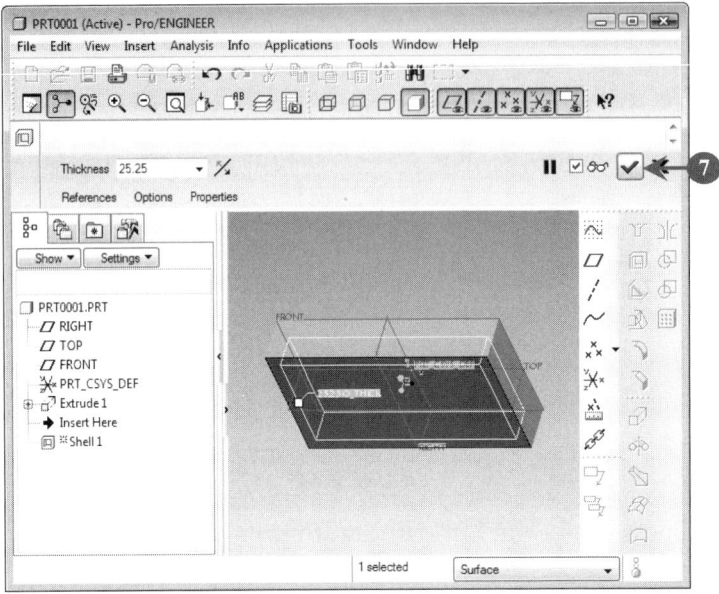

FIGURE 4.66

The shell appears, as shown in **Figure 4.67**:

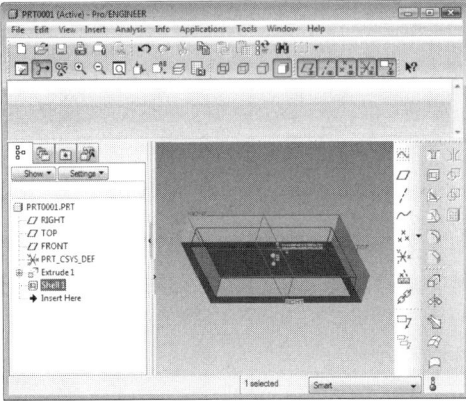

FIGURE 4.67

The boundary of the shell will be highlighted in red on your computer (Figure 4.67).

Note: As soon as the **Build Feature** button is clicked the **Shell** dashboard disappears (Figure 4.67).

---

8. *Click* anywhere in the **Drawing Area**. The final view of the shell appears, as shown in **Figure 4.68**:

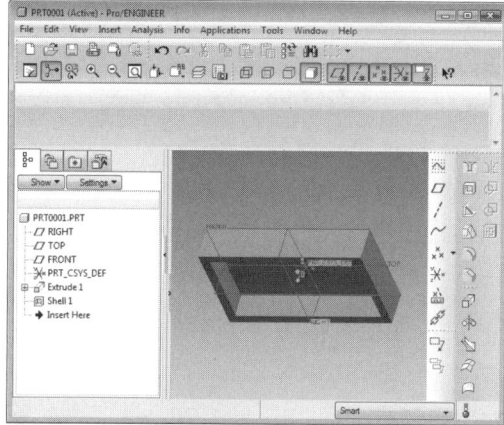

FIGURE 4.68

9. Finally, *save* the file. In our case, the file is saved with the name SHELLFEATURE.prt.

Here, we have completed the process of creating a shell on a base feature. Next, we move ahead and understand the process of creating a revolved feature.

4.9 INDUCING A REVOLVED FEATURE

The revolved feature in Pro/ENGINEER Wildfire 4.0 is used to create features by revolving a sketch around the centerline. To create a revolved feature the **Revolve** dashboard is used, which helps you create features such as revolved protrusions, revolved cuts, and revolved surfaces. However, to create a revolved feature, a base feature is required. Therefore, in this section, we use the PRT0001.prt file as the base feature, which contains a rectangular slot. On that rectangular slot, we create a revolved protrusion. Let's follow the steps given here to create a revolved protrusion:

1. *Start* the **Part** mode and open the PRT0001.prt file (Figure 4.24).
2. *Click* the **Revolve** (⬠) button in the **Base Features toolbar** (Figure 4.11) in the **right tool chest**. The **Revolve** dashboard appears (**Figure 4.69**).

3. *Select* the **Placement** tab in the **Revolve** dashboard, as shown in Figure 4.69:

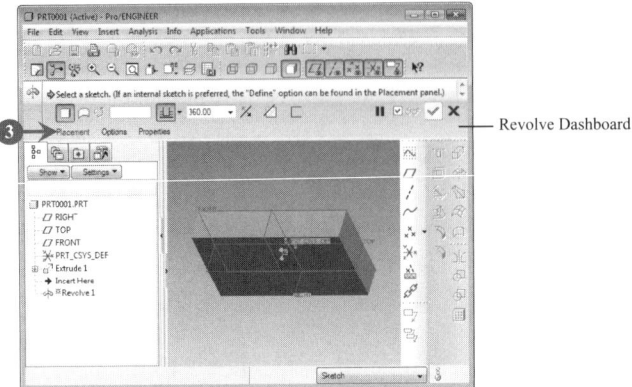

FIGURE 4.69

A slide-down panel appears that lets you define the sketching plane (**Figure 4.70**).

4. *Click* the **Define** button on the slide-down panel, as shown in Figure 4.70:

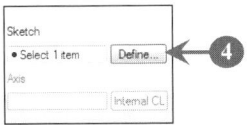

FIGURE 4.70

The **Sketch dialog box** where you have to specify the sketching plane, its reference plane, and orientation appears (Figure 4.14).

5. *Select* the top surface of the rectangular slot (Figure 4.69). The selected surface, its reference plane, and orientation appear in the **Sketch dialog box** (**Figure 4.71**).
6. *Click* the **Sketch** button, as shown in Figure 4.71:

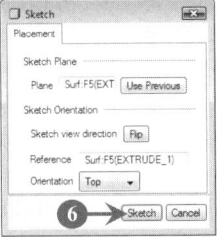

FIGURE 4.71

The **Sketcher tools toolbar** appears to let you draw the sketch of the protrusion, as shown in **Figure 4.72**:

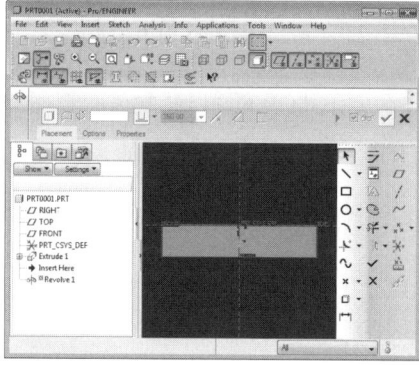

FIGURE 4.72

7. Now, *draw* a sketch. In our case, we have drawn a circle, as shown in **Figure 4.73**.

Note: The process to sketch a circle is depicted in Chapter 3, "Exploring Pro/ENGINEER Wildfire 4.0 Sketch Mode."

8. *Draw* a centerline around the circle to use it as an axis of revolution (Figure 4.73).

Note: The process to sketch a centerline is depicted in Chapter 3, "Exploring Pro/ENGINEER Wildfire 4.0 Sketch Mode."

9. *Click* the **Done** (✔) button, as shown in Figure 4.73:

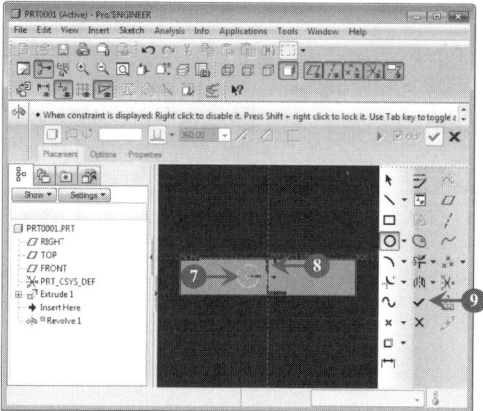

FIGURE 4.73

The sketch of the revolved protrusion is created, as shown in **Figure 4.74**.

> **Note:** The revolved protrusion is created according to the default setting of the **Revolved** dashboard. For example, the default angle for the protrusion is 360°; however, if you want to modify the angle, then click the down arrow beside ⌗360.00 ▾⌗ and select another angle. You can also enter an angle rather than selecting an angle from the drop-down edit list.

10. *Click* the **Preview** (⌗☑∞⌗) button to verify the revolved protrusion, as shown in Figure 4.74:

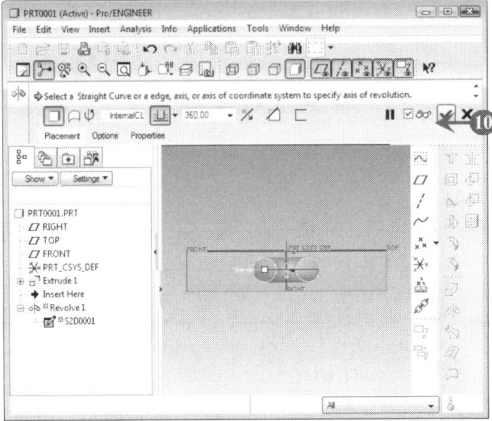

FIGURE 4.74

The preview of the revolved protrusion appears, as shown in **Figure 4.75**:

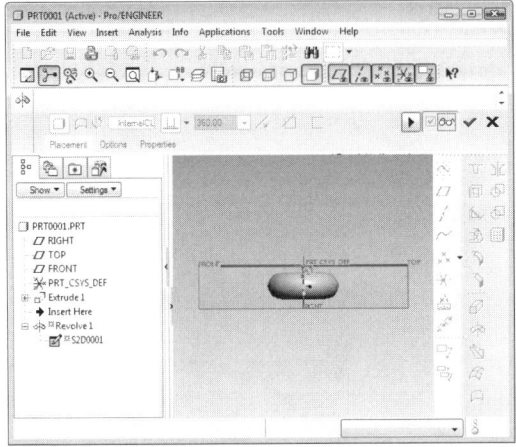

FIGURE 4.75

11. *Hold* the middle mouse button and spin the rectangular slot to have a 3D view of the protrusion, as shown in **Figure 4.76**.

12. *Click* the **Build feature** (☑) button to complete the revolved protrusion, as shown in Figure 4.76:

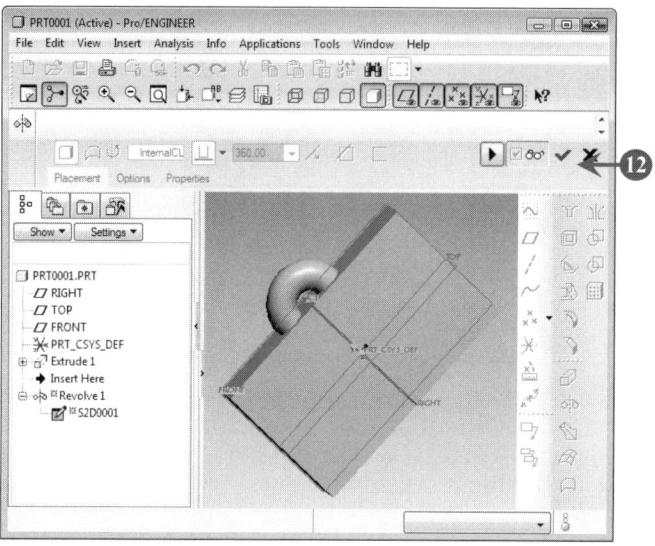

FIGURE 4.76

The revolved protrusion is highlighted, as shown in **Figure 4.77**:

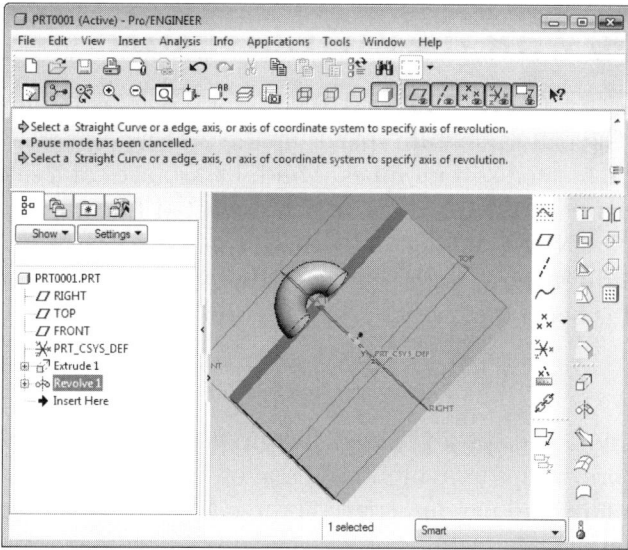

FIGURE 4.77

13. *Click* anywhere in the **Drawing Area**. The final view of the revolved protrusion is shown in **Figure 4.78**:

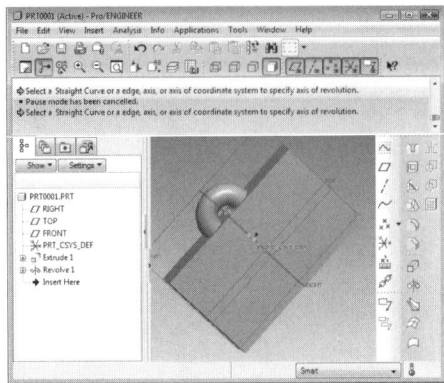

FIGURE 4.78

14. Finally, *save* the file.

This is the process of creating a revolved protrusion. Next, we study the process to create multiple copies of a feature.

4.10 WORKING WITH THE PATTERN FEATURE

The pattern feature in Pro/ENGINEER Wildfire 4.0 creates multiple instances of a selected feature. The pattern feature speeds up the process of model creation where a feature needs to be created several times. Multiple instances of a selected feature are called the **patterns** or **child features** and the selected feature is called the **parent** feature. In Pro/ENGINEER Wildfire 4.0 various types of patterns such as **Dimension** pattern or **Direction** pattern can be created. However, before understanding pattern types, we discuss the benefits of creating patterns. Following are the benefits of creating patterns:

- Multiple copies of a selected feature can be created, which saves time; therefore, the process of model creation speeds up.
- All patterns are parameterized and modifying the parameters, such as the number of child features or spacing between child features is easy.
- All patterns work as a single feature; therefore, modifying the dimensions of all the copies is easy because if you modify the dimension of the parent feature, then the change is reflected in all the child features.

Let's see **Table 4.1**, which lists the type of patterns that can be created in Pro/ENGINEER Wildfire 4.0:

Pattern Type	Explanation
Dimension	Creates the child features using the dimensions of the parent feature.
Direction	Creates the child features by specifying a direction on the base feature. The direction can be specified by selecting a sketching plane or an edge.
Axis	Creates the child features by specifying an angle around an axis. The axis should be of the feature around which the child features have to be created.
Table	Creates the child features by using a pattern table. In the pattern table, you need to specify the dimension of every child feature.
Reference	Creates the child features by referencing the parent feature.
Fill	Creates the child features by filling the sketched area with a selected feature.
Curve	Creates the child features by specifying the numbers of the child features along a curve.

TABLE 4.1   Pattern types available in Pro/ENGINEER Wildfire 4.0

There are the various types of patterns that can be created in Pro/ENGINEER Wildfire 4.0. However, in this section, we see the process to create the child features using the **Axis** pattern in the following steps:

1. *Start* the **Part** mode and *open* the HOLEFEATURE.prt file (Figure 4.42).
2. *Create* a hole of diameter 30.00 adjacent to the hole in the base feature. Once the hole has been created, the base feature appears (**Figure 4.79**).
3. Now, *select* the hole whose patterns need to be created, as shown in Figure 4.79:

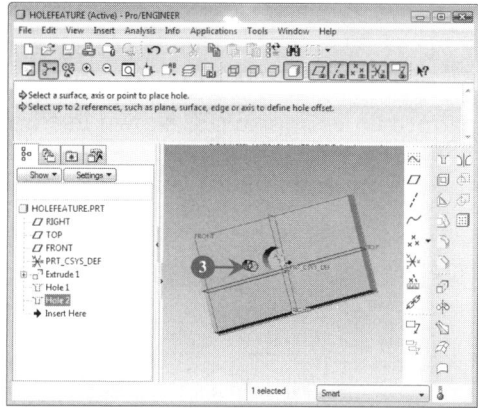

FIGURE 4.79

In our case, the hole on the left side (Figure 4.79) is selected.

4. *Click* the **Pattern** (▥) button in the **Edit Features toolbar**, as shown in **Figure 4.80**:

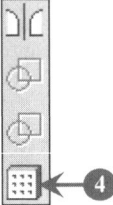

FIGURE 4.80

Note: By default, all buttons in the **Edit Feature toolbar** are disabled. These buttons are enabled depending on their requirements. For example, the **Pattern** button is enabled when a feature that is to be patterned is selected.

The **Pattern** dashboard appears (**Figure 4.81**).

5. *Click* the down-arrow beside ▤Dimension ▾, as shown in Figure 4.81:

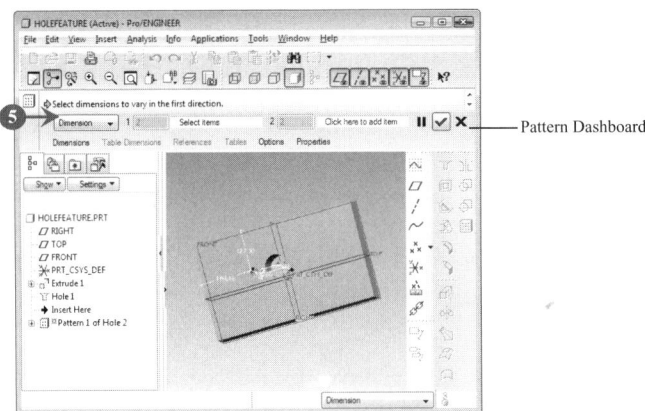

FIGURE 4.81

A list of various pattern types appears, as shown in **Figure 4.82**.

6. *Select* the **Axis** pattern from the list (Figure 4.82):

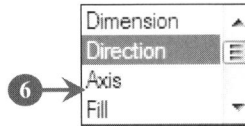

FIGURE 4.82

The layout of the **Pattern** dashboard changes (**Figure 4.83**).

7. *Select* the axis, A_1, of the hole, as shown in Figure 4.83:

Pattern Dashboard
with Changed Layout

FIGURE 4.83

Note: The layout of the **Pattern** dashboard depends on the type of pattern selected.

The selected axis turns red.

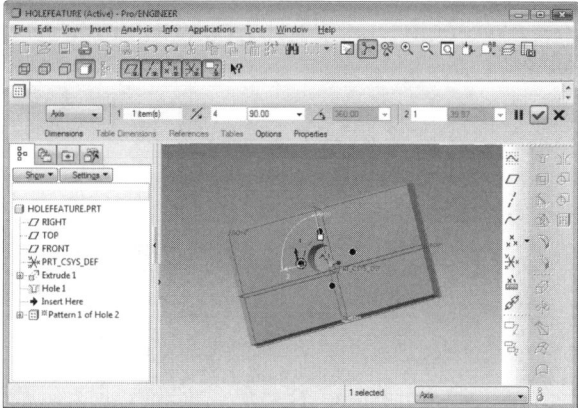

FIGURE 4.84

Apart from that, a preview of the patterns also appears (**Figure 4.84**). You see four holes around the selected axis (Figure 4.84); this is because by default, the number of patterns are four and the angle between these patterns is 90°.

8. To modify the total number of patterns, *specify* a new number in the text box beside the **Flip** (⚡) button. In our case, we have entered 6.

Note: The **Flip** button is used to flip the angular direction of the pattern.

9. To modify the angle between the patterns, either *select* an angle from the list of predefined angles by *clicking* the down arrow beside ⬚ or enter any other angle. In our case, we have selected 60° from the list of given angles.

10. *Press* the **ENTER** key. Changes are reflected in the **Drawing Area**, as shown in **Figure 4.85**.

11. *Click* the **Build Feature** (⬚) button, as shown in Figure 4.85:

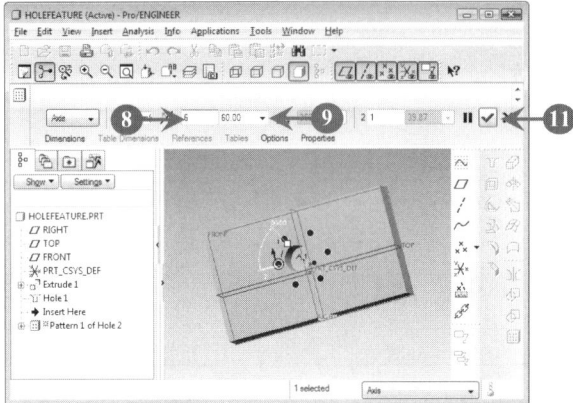

FIGURE 4.85

All the patterns are highlighted in red, as shown in **Figure 4.86**:

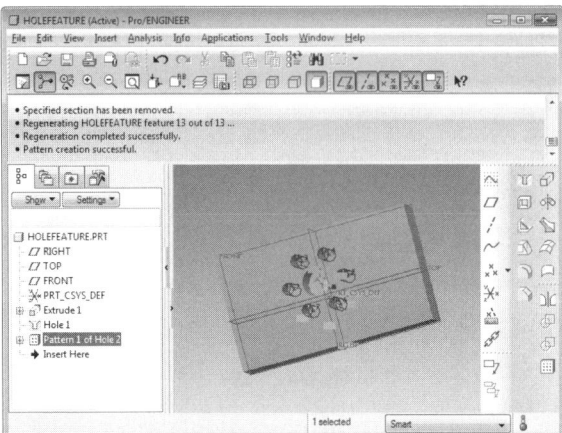

FIGURE 4.86

12. *Click* anywhere in the **Drawing Area** to have a look at the final view of the patterns. The base feature along with the patterns appears, as shown in **Figure 4.87**:

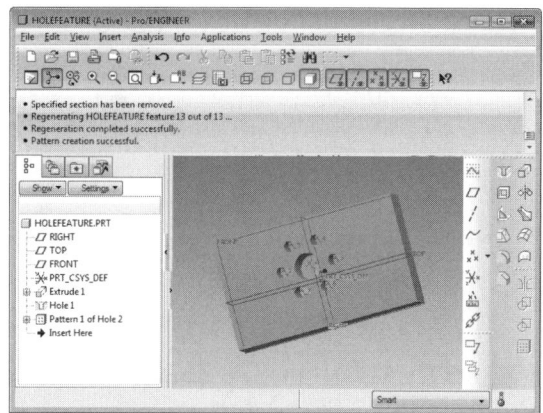

FIGURE 4.87

13. *Save* the file. We have saved the file with the name PATTERNFEATURE.prt.

We have learned the process to create **Axis** patterns in Pro/ENGINEER Wildfire 4.0. Next, we study a feature called **Sweep**.

4.11 INDUCING THE SWEEP FEATURE

A sweep feature in Pro/ENGINEER Wildfire 4.0 is created by extruding a section along a trajectory. The trajectory can be sketched during a feature creation or can be selected within an already created feature. In this section, we see the process to create a sweep feature through sketching a section. Let's follow the steps given here to create a sweep feature:

1. *Start* the **Part** mode. *Select* **Insert > Sweep > Protrusion**, as shown in **Figure 4.88**:

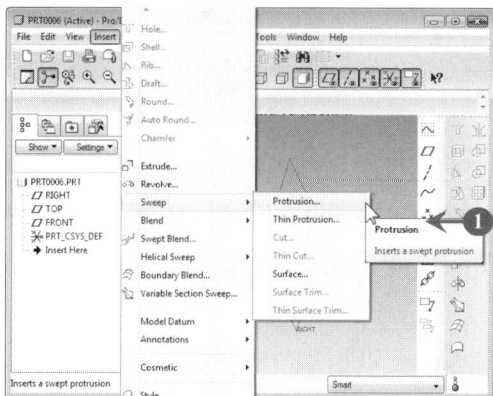

FIGURE 4.88

The **PROTRUSION: Sweep dialog box** appears, as shown in **Figure 4.89**:

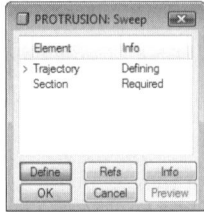

FIGURE 4.89

Apart from the **PROTRUSION: Sweep dialog box**, a **Menu Manager menu** also appears (**Figure 4.90**).

The **Menu Manager menu** provides you with two options:

- To sketch a trajectory.
- To select a trajectory from the existing feature.

2. *Select* the **Sketch Traj** option, as shown in Figure 4.90:

FIGURE 4.90

The appearance of the **Menu Manager menu** changes, as shown in **Figure 4.91**:

FIGURE 4.91

Apart from the **Menu Manager menu**, a **Select dialog box** also appears, as shown in **Figure 4.92**:

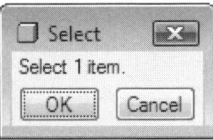

FIGURE 4.92

The **Select dialog box** prompts you to select an item; that is, select the datum plane to define a trajectory along the selected datum plane.

3. *Select* the **Front** datum plane either in the **Drawing Area** or in the **Model Tree** in the **Part** mode window. A red arrow appears in the direction of viewing the selected datum plane, as shown in **Figure 4.93**:

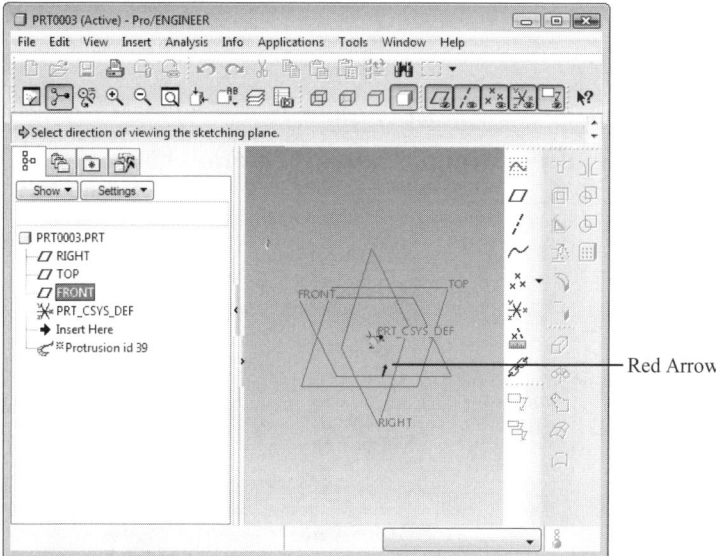

FIGURE 4.93

Apart from this, the appearance of the **Menu Manager menu** changes once again (**Figure 4.94**).

4. *Select* the **Okay** option to continue creating a sweep feature, as shown in Figure 4.94:

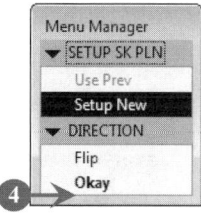

FIGURE 4.94

The appearance of the **Menu Manager menu** changes again (**Figure 4.95**). Now, you have to select the orientation view of the sweep feature from the options given in the **SKET VIEW** submenu of the **Menu Manager menu** (Figure 4.95).

5. *Select* the **Top** option in the **SKET VIEW** submenu, as shown in Figure 4.95:

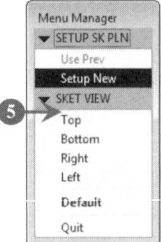

FIGURE 4.95

The appearance of the **Menu Manager menu** changes again, as shown in **Figure 4.96**:

FIGURE 4.96

Apart from this, the **Select dialog box** (Figure 4.92) appears. Both the **Menu Manager menu** (Figure 4.96) and the **Select dialog box** prompt you to select the reference plane of the **Front** datum plane in the **Part** mode window.

6. *Select* the **Top** datum plane in the **Model Tree**. The **Sketcher tools toolbar** appears to let you sketch the trajectory (**Figure 4.97**).
7. *Select* the **3 Point/Tangent End** (⟍•) button in the **Sketcher tools toolbar** (Figure 4.97) to draw an arc, as shown in Figure 4.97:

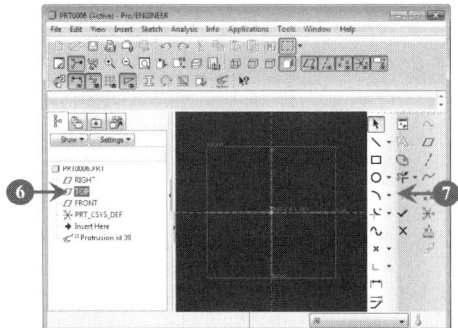

FIGURE 4.97

A yellow arrow is attached to the arc (**Figure 4.98**). This yellow arrow represents the starting point of the trajectory and it points in the direction of the sweep.

Note: The process of drawing an arc is depicted in Chapter 3, "Exploring Pro/ENGINEER Wildfire 4.0 Sketch Mode."

---

8. *Click* the **Done** (☑) button to complete the sketch of the trajectory, as shown in Figure 4.98:

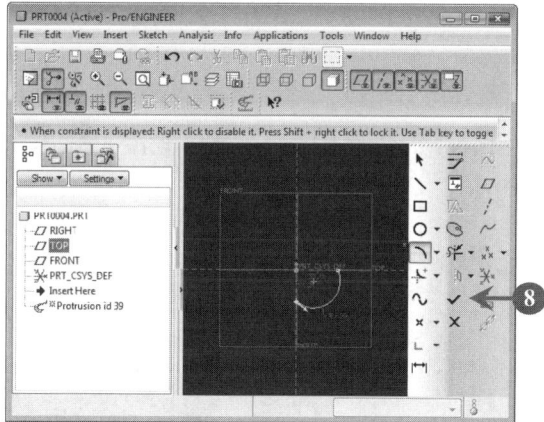

FIGURE 4.98

The view of the **Drawing** mode is changed. This lets you create a section for sweep, as shown in **Figure 4.99**. The intersection point of two infinite yellow lines represents the starting point of the trajectory. The starting point of the section, which is to be created for a sweep, should lie at the starting point of the trajectory.

9. *Click* the **Rectangle** (☑) button (Figure 4.99) in the **Sketcher tools toolbar** to draw a rectangle as a section for the sweep.

Note: The process to draw a rectangle is depicted in Chapter 3, "Exploring Pro/ENGINEER Wildfire 4.0 Sketch Mode."

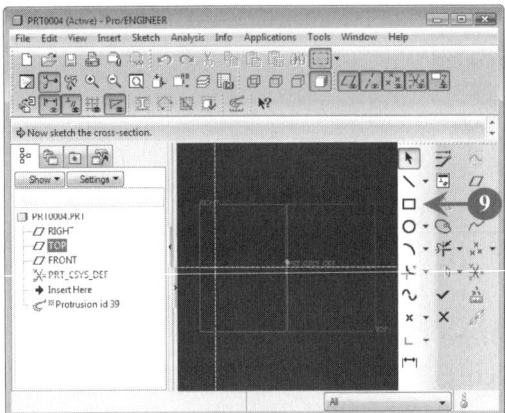

FIGURE 4.99

The rectangle appears in yellow (**Figure 4.100**).

10. *Click* the **Done** (☑) button in the **Sketcher tools toolbar**, as shown in Figure 4.100:

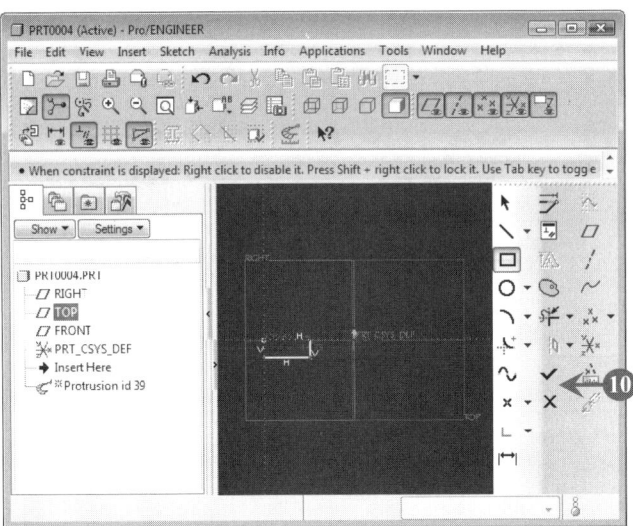

FIGURE 4.100

The sweep feature is created, as shown in **Figure 4.101**:

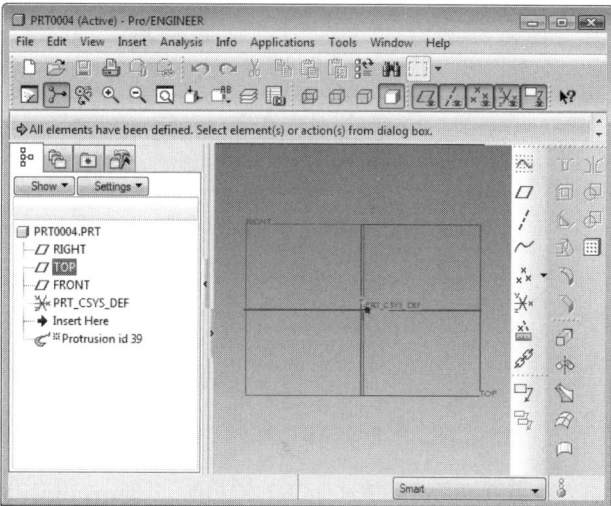

FIGURE 4.101

In Figure 4.101, the datum planes, along with the sweep feature, are displayed.

11. *Click* the **Preview** button in the **PROTRUSION: Sweep dialog box** (Figure 4.89) to view the solid sweep feature. The sweep feature appears, as shown in **Figure 4.102**:

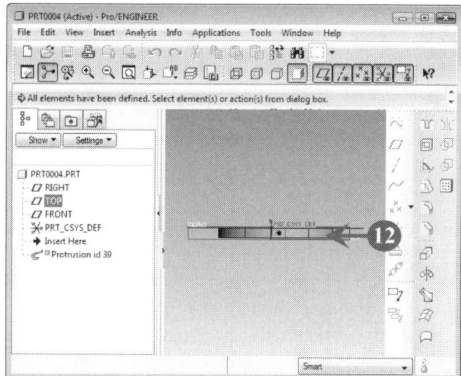

FIGURE 4.102

12. *Select* the sweep feature (Figure 4.102), *hold* the middle mouse button, and *spin* the sweep feature to view it from different angles, as shown in **Figure 4.103**:

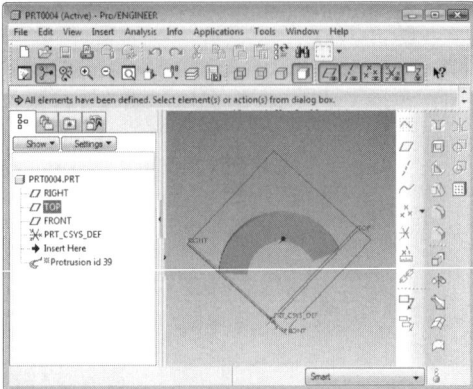

FIGURE 4.103

13. *Click* the **OK** button in the **PROTRUSION: Sweep dialog box** (Figure 4.89) to finish the process. The sweep feature is highlighted in red, as shown in **Figure 4.104**:

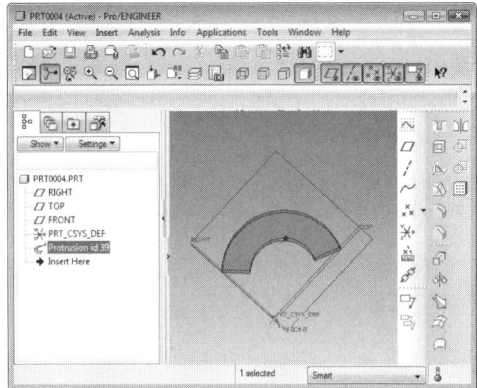

FIGURE 4.104

14. *Save* the file.

Presently, we studied the process to create a sweep feature. Next, we study another feature, which blends two sections of different features to create a single feature.

4.12 INDUCING THE BLEND FEATURE

A blended feature in Pro/ENGINEER Wildfire 4.0 is created by joining two or more sections at their edges through transitional faces. The blending of two or more sections results in a continuous feature. However, an important point to remember while creating a blended feature is that the total number of entities in all of the sections must be the same. For example, a rectangle has four entities (four lines); therefore, it can either be blended with another rectangle or with any section having four entities. In Pro/ENGINEER Wildfire 4.0, different blend types are used to blend different sections. The different blend types are

- **Parallel:** This type is used to blend the sections that are drawn parallel to each other.
- **Rotational:** This type is used to blend the sections that are rotated about the Y-axis up to an angle of maximum 120°. The angle between the sections is known as the rotational blend angle. If the angle between the sections is 0°, then this option works as the **Parallel** blend option. All of the sections that are blended using this option are sketched individually and aligned using the coordinate system of the sections.
- **General:** This type is used to blend the sections that can be rotated and translated on the *x*-, *y*-, and *z*- axes. All of the sections blended using this option are also sketched individually and aligned using the coordinate system of the sections.

In this section, we use the **Parallel** blend option to blend two rectangular sections. Following are the steps that help you to blend two rectangular sections:

1. *Start* the **Part** mode.
2. *Select* **Insert > Blend > Protrusion**, as shown in **Figure 4.105**:

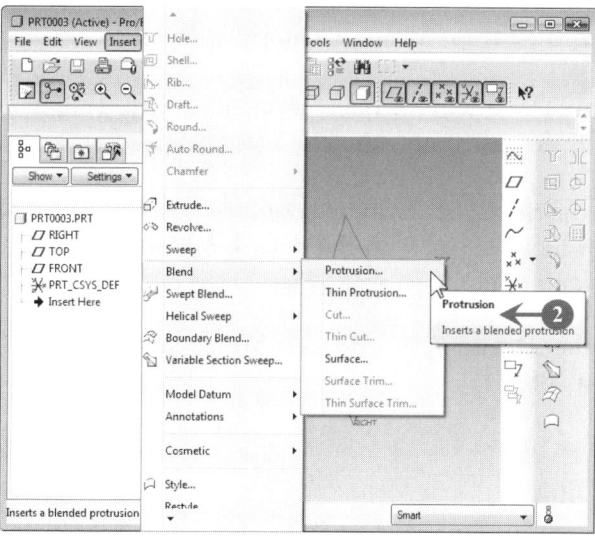

FIGURE 4.105

The **Menu Manager menu** appears, as shown in **Figure 4.106**.

3.   *Select* **Done** in the **Menu Manager menu**, as shown in Figure 4.106:

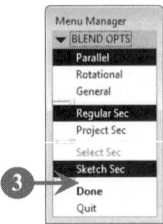

FIGURE 4.106

The **PROTRUSION: Blend**, **Parallel**, **Regular Sections dialog box** appears, as shown in **Figure 4.107**:

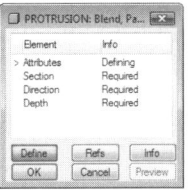

FIGURE 4.107

Apart from the **PROTRUSION: Blend**, **Parallel**, **Regular Sections dialog box**, the appearance of the **Menu Manager menu** has changed (**Figure 4.108**).
    This time the **Menu Manager menu** appears with the following attributes:

- **Straight:** This attribute joins the sections with straight lines
- **Smooth:** This attribute joins the sections with curves

By default, the **Straight** attribute is selected. Now, either select any one of the attributes or move ahead with the default option. In our case, we move ahead with the default option.

4.   *Select* the **Done** option in the **Menu Manager menu**, as shown in Figure 4.108:

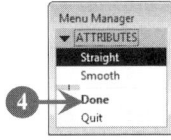

FIGURE 4.108

The appearance of the **Menu Manager menu** changes, as shown in **Figure 4.109**:

FIGURE 4.109

Apart from the **Menu Manager menu** (**Figure 4.109**), the **Select dialog box** also appears, as shown in **Figure 4.110**:

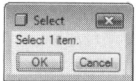

FIGURE 4.110

Both the **Menu Manager menu** (Figure 4.109) and the **Select dialog box** (Figure 4.110) prompt you to select the datum plane along with the sections that are to be drawn.

5. *Select* the **Front** datum plane either in the **Model Tree** or in the **Drawing Area**. A red arrow appears on the **Front** datum plane, which points in the direction of feature creation, as shown in **Figure 4.111**:

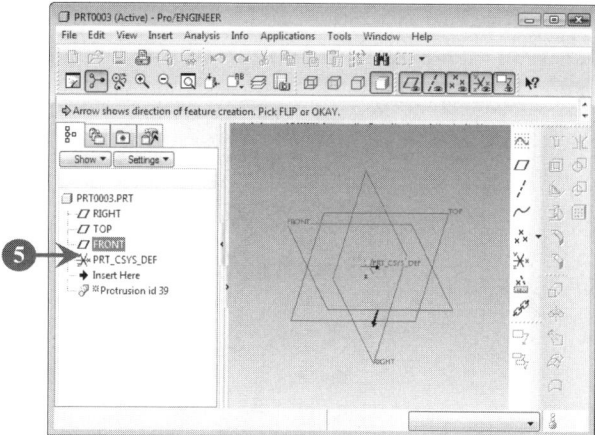

FIGURE 4.111

Apart from this, the appearance of the **Menu Manager menu** changes once again, as shown in **Figure 4.112**.

Now, either flip the direction of the feature creation by selecting the **Flip** option in the **Menu Manager menu** (Figure 4.112) or accept the default direction by selecting the **Okay** option. In our case, we *select* the **Okay** option.

6.   *Select* the **Okay** option in the **Menu Manager** menu, as shown in Figure 4.112:

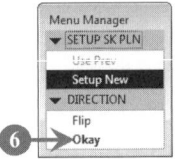

FIGURE 4.112

The appearance of the **Menu Manager menu** changes once again (**Figure 4.113**). Now, you need to select any one of the options given in the **SKET VIEW** submenu of the **Menu Manager menu** to set the orientation of the section.

7.   In our case, we *select* the **Top** view, as shown in Figure 4.113:

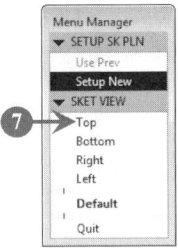

FIGURE 4.113

The appearance of the **Menu Manager menu** changes again, as shown in **Figure 4.114**:

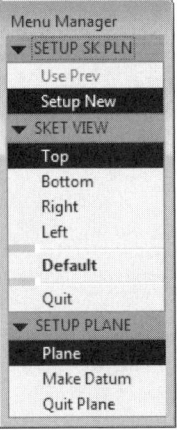

FIGURE 4.114

Apart from this, the **Select dialog box** appears (Figure 4.110). Both the **Menu Manager menu** (Figure 4.114) and the **Select dialog box** prompt you to select the **Reference** datum plane of the **Front** datum plane.

8. *Select* the **Top** datum plane (**Figure 4.115**) in the **Model Tree** as the **Reference** datum plane. The **Sketcher tools toolbar** appears, as shown in Figure 4.115:

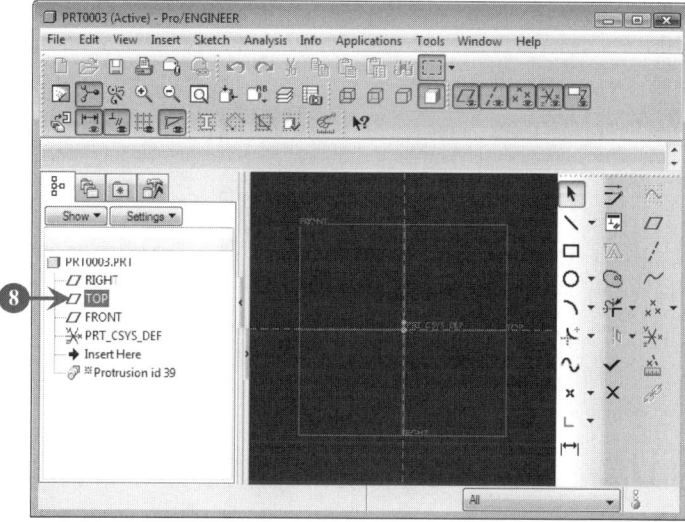

FIGURE 4.115

9. Now, draw a rectangle, as shown in **Figure 4.116**. A yellow arrow is attached to the rectangle (Figure 4.116). This yellow arrow shows the starting point of the rectangle.
10. *Click* the **Done** (✓) button in the **Sketcher tools toolbar**, as shown in Figure 4.116:

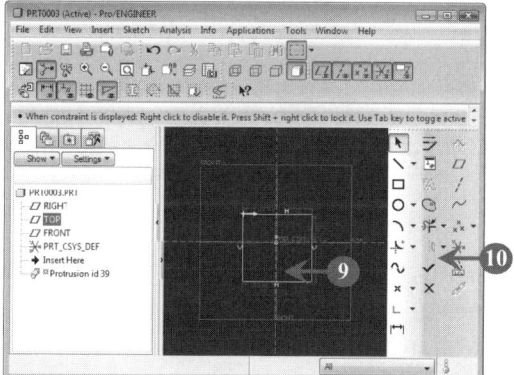

FIGURE 4.116

Now, create another section. However, to create another section, you must invoke the **Toggle Section** option from the **Sketch menu**. If the **Toggle Section** option is not invoked, then another section cannot be created.

11. *Select* **Sketch**>**Feature Tools**>**Toggle Section**, as shown in **Figure 4.117**:

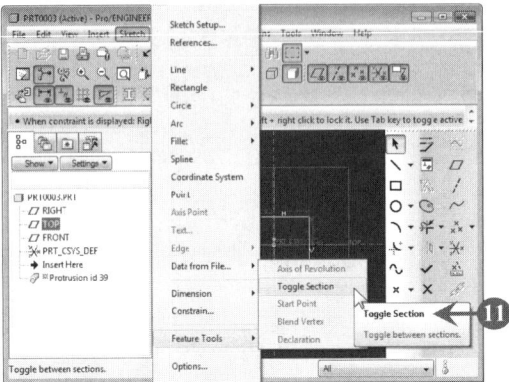

FIGURE 4.117

All inactive portions of the section turn gray, as shown in **Figure 4.118**:

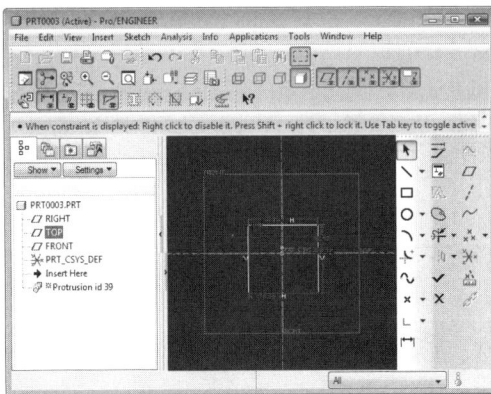

FIGURE 4.118

12. Now, *create* another rectangle, as shown in **Figure 4.119**.
13. *Click* the **Done** () button in the **Sketcher tools toolbar** to complete the section, as shown in Figure 4.119:

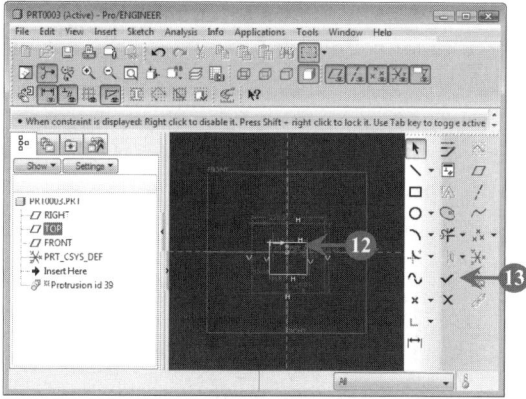

FIGURE 4.119

A text box that allows you to Enter **DEPTH** for section 2 appears above the **Drawing Area**, as shown in **Figure 4.120**.

14. Now, either specify a depth value for section 2 or continue with the default value appearing in the text box (Figure 4.120). We are continuing with the default value.

15. *Click* the **Accept Value** ( ) button to continue, as shown in Figure 4.120:

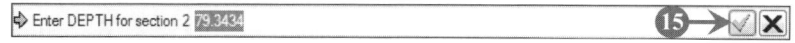

FIGURE 4.120

Both sections are now blended, as shown in **Figure 4.121**:

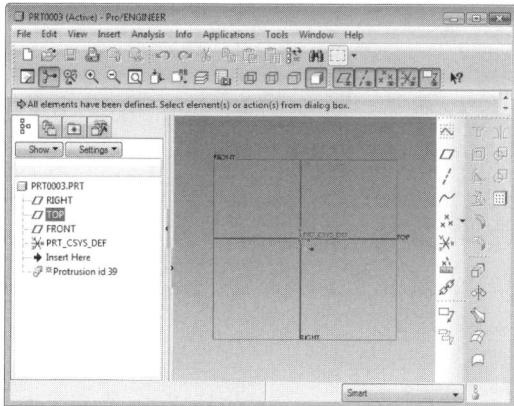

FIGURE 4.121

In Figure 4.121, you can view the datum planes along with the sections that are blended. To have a look on the solid display of the blended sections, you need to preview the blended sections.

16. *Click* the **Preview** button in the **PROTRUSION: Blend, Parallel, Regular Sections dialog box** (Figure 4.107). The preview of the blended sections appears, as shown in **Figure 4.122**:

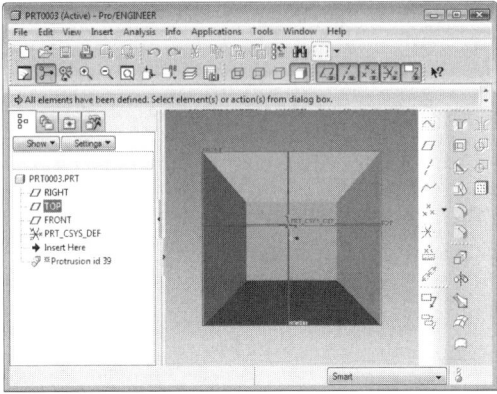

FIGURE 4.122

17. *Click* the **OK** button in the **PROTRUSION: Blend**, **Parallel**, **Regular Sections dialog box** (Figure 4.107) to finish the process. The blended sections appear in red, as shown in **Figure 4.123**:

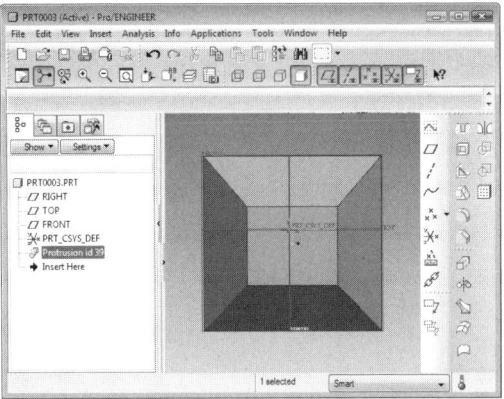

FIGURE 4.123

18. Finally, *save* the file.

With this, we complete the study of the blending feature. Next, we study another feature called the **Rib** feature, which is used to provide support to the features in a model.

4.13 INDUCING THE RIB FEATURE

A **Rib** is used to support and strengthen the weak features of a model. A rib in Pro/ENGINEER Wildfire 4.0 is created using the **Rib** dashboard. However, a rib is created along with a feature; therefore, we first need a feature to create a rib. In this section, we follow the rib creation after the feature creation. Let's follow the steps given here to create a rib:

1. *Start* the **Part** mode and *create* a feature, as shown in **Figure 4.124**.

The feature shown in Figure 4.124 is created by drawing a horizontal rectangular slot and a vertical rectangular slot.

> **Note:** The process to create a rectangular slot is similar to creating a base feature, which is discussed in the section "Creating a Base Feature."

2. Next, *cli*ck the **Rib** button (⬛) in the **Engineering Features toolbar** to invoke the **Rib** dashboard, as shown in Figure 4.124:

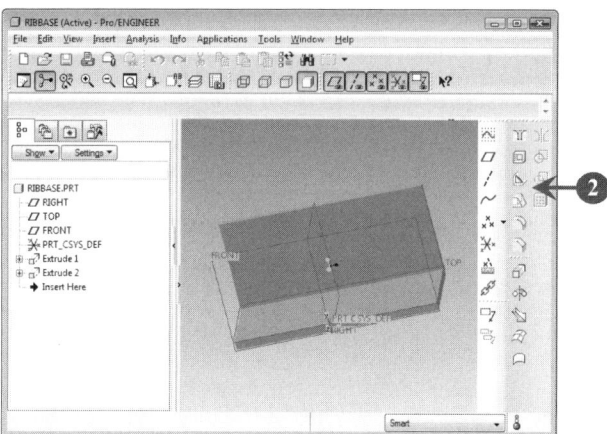

FIGURE 4.124

The **Rib** dashboard appears (**Figure 4.125**).

3.   *Select* the **References** tab in the **Rib** dashboard, as shown in Figure 4.125:

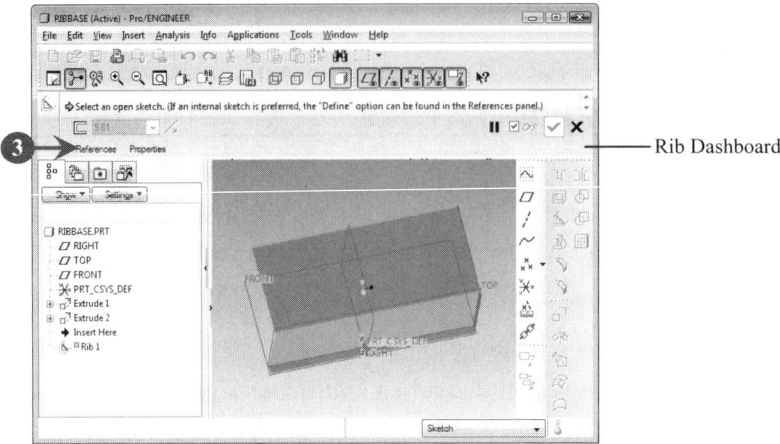

FIGURE 4.125

A slide-down panel that helps select the sketch plane to draw the sketch of the rib appears (**Figure 4.126**).

4.   *Click* the **Define** button in the slide-down panel, as shown in Figure 4.126:

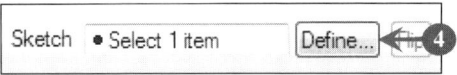

FIGURE 4.126

The **Sketch dialog box** appears (**Figure 4.127**):

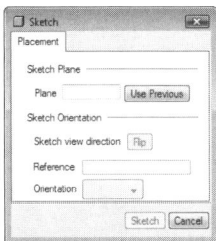

FIGURE 4.127

5.   *Select* the **Right** datum plane, either in the **Model Tree** or in the **Drawing Area.** The **Right** datum plane and its **Reference** datum plane, along

with the orientation, automatically appear in the **Sketch dialog box** (**Figure 4.128**).

> **Note:** *Select* the **Orientation** as **Top**. *Select* the **Orientation** by clicking the down arrow beside the **Orientation** drop-down list.

6.   *Click* the **Sketch** button, as shown in Figure 4.128:

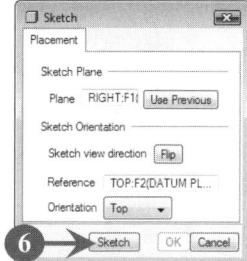

FIGURE 4.128

The **Sketcher tools toolbar** appears to let you create the sketch of the rib, as shown in **Figure 4.129**:

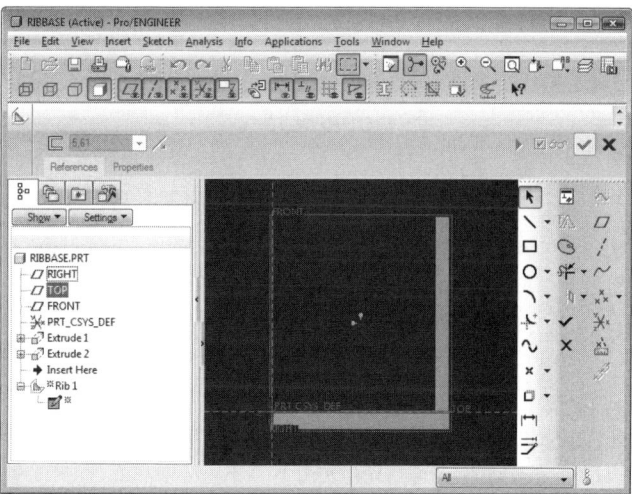

FIGURE 4.129

Now, you have to set the reference line along which the rib will be created.

7. *Select* the **References** option in the **Sketch menu**, as shown in **Figure 4.130**:

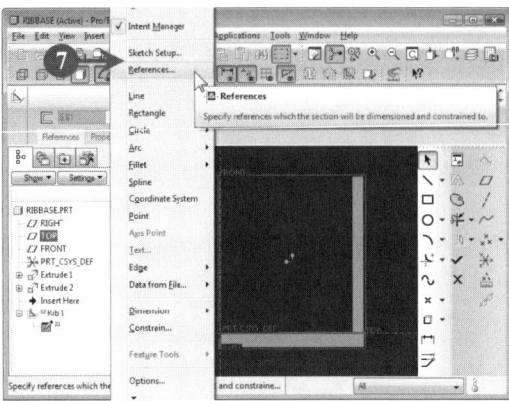

FIGURE 4.130

As soon as the **References** option is selected, the **References** window appears, as shown in **Figure 4.131**:

FIGURE 4.131

Apart from that, the **Select dialog box** also appears, as shown in **Figure 4.132**:

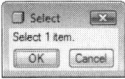

FIGURE 4.132

The **Select dialog box** prompts you to select an item; that is, select the surface along with the reference line that will be created.

8. *Select* the surface along with the reference line to be created. An infinite reference line appears, as shown in **Figure 4.133**:

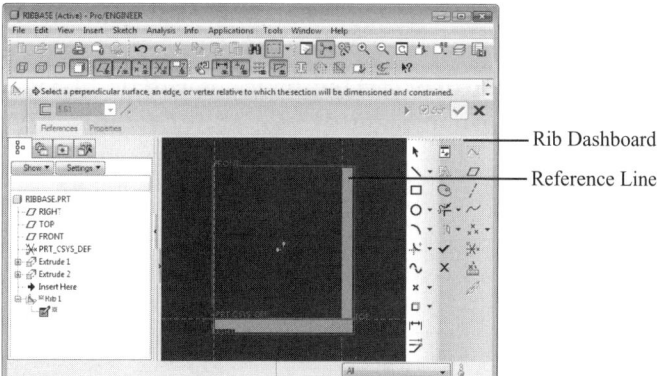

FIGURE 4.133

The selected surface is automatically added in the **References** window.

9. *Select* the cross section of the two infinite lines along with the rib to be created.
10. Next, *draw* the sketch of the rib (**Figure 4.134**).
11. *Click* the **Done** (☑) button to complete the sketch and to exit from the **Sketcher tools toolbar** (Figure 4.134):

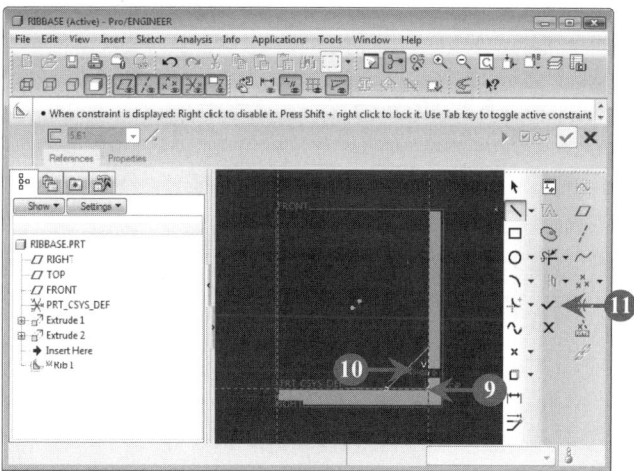

FIGURE 4.134

The rib is previewed, as shown in **Figure 4.135**.

12. *Enter* 8.50 in the **Dimension** (5.61 ▾) **Box** in the **Rib** dashboard (Figure 4.135).

13. *Press* the **ENTER** key to project the new value in the **Dimension Box** (Figure 4.135).

14. *Click* the **Build Feature** (☑) button in the **Rib** dashboard to complete the rib creation, as shown in Figure 4.135:

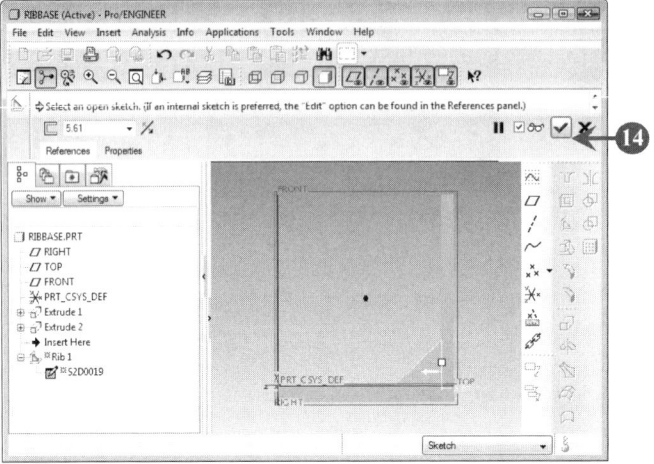

FIGURE 4.135

The rib is highlighted in red, as shown in **Figure 4.136**:

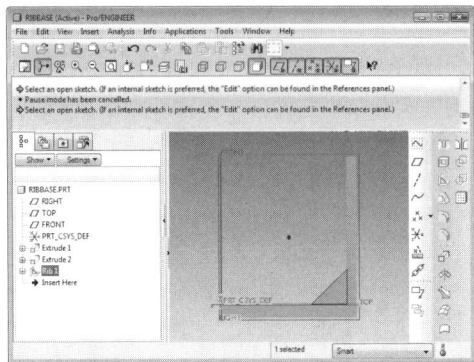

FIGURE 4.136

15. Finally, *save* the file.

This is the process to create a rib in Pro/ENGINEER Wildfire 4.0. We have also completed creating and inducing features in Pro/ENGINEER Wildfire 4.0. Next, we see the process to modify features.

4.14 MODIFYING FEATURES

Sometimes during model creation or after creating a feature, you may need to modify a feature to get the desired feature or model. A feature in Pro/ENGINEER Wildfire 4.0 can be modified by:

- Modifying Dimensions
- Modifying Definition

Let's first study modifying dimensions.

Modifying Dimensions

Features are generally modified by editing their dimensions. However, to learn the process of modifying a feature by editing its dimensions, we need a feature. Therefore, we use the PRT0001.prt file, which we created previously and contains a rectangular slot (Figure 4.24). Let's now follow the steps given here to understand the process of modifying dimensions:

1. *Start* the **Part** mode and *open* the PRT0001.prt file (Figure 4.24).
2. *Select* the feature whose dimensions need to be modified in the **Model Tree**. In our case, we have selected the feature named **Extrude 1**. The selected feature is highlighted in red, as shown in **Figure 4.137**:

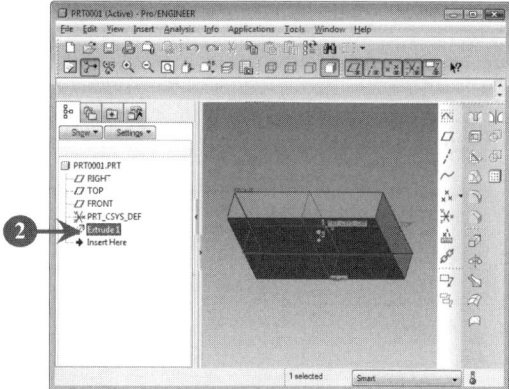

FIGURE 4.137

3. *Right-click* the selected feature in the **Model Tree** (Figure 4.137). A context menu appears (**Figure 4.138**).
4. *Select* the **Edit** option in the **Context menu**, as shown in Figure 4.138:

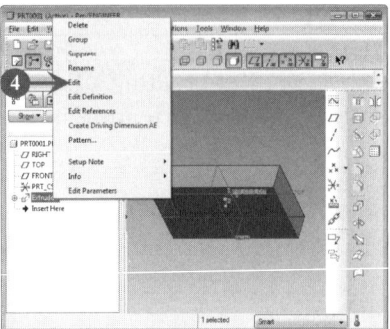

FIGURE 4.138

All the dimensions of the selected feature are highlighted in yellow, as shown in **Figure 4.139**:

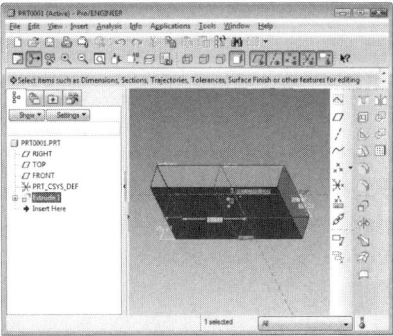

FIGURE 4.139

5. *Select* the dimension you want to modify. The selected dimension turns red, as shown in **Figure 4.140**:

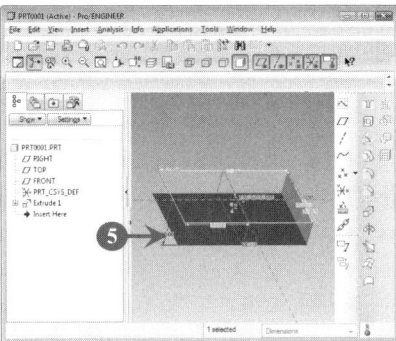

FIGURE 4.140

6. *Double-click* the dimension (Figure 4.140). An **Edit Box** appears, as shown in **Figure 4.141**:

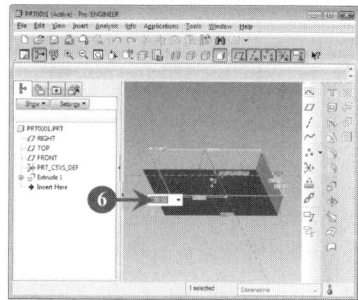

FIGURE 4.141

7. *Specify* a new dimension in the **Edit Box** (Figure 4.141).
8. After specifying the new dimension, *press* the **ENTER** key. The new dimension is highlighted in light green (**Figure 4.142**).
9. *Click* the **Regenerate** (![icon]) button that is placed on the **top tool chest**, as shown in Figure 4.142:

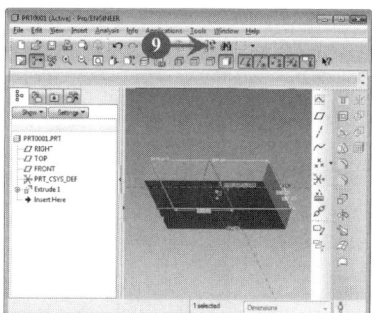

FIGURE 4.142

The selected feature is regenerated according to the new dimensions, as shown in **Figure 4.143**:

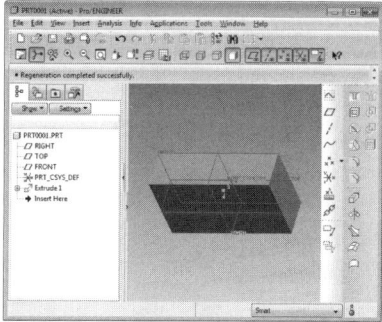

FIGURE 4.143

10. Finally, *save* the modified file.

Modifying Definition

Features can also be modified by editing their definition. Editing the definition of a feature refers to modifying the parameters used to create that feature. In this section, we use the PATTERNFEATURE.prt file to understand the process of modifying a feature by editing its definition. This file contains a pattern of six holes around another hole on a rectangular slot. We will modify this pattern by editing its definition. Let's follow the steps given here to understand the process of modifying the definition of a feature:

1. *Start* the **Part** mode and *open* the PATTERNFEATURE.prt file (Figure 4.87).
2. *Select* the feature whose definition needs to be modified in the **Model Tree**. In our case, we have selected the pattern feature named **Pattern 1 of Hole 2** (Figure 4.87). The selected feature is highlighted in red, as shown in **Figure 4.144**:

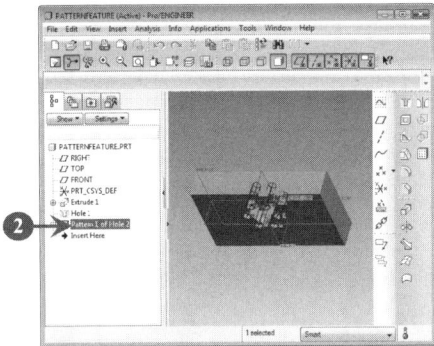

FIGURE 4.144

3. *Right-click* the selected feature (Figure 4.144). A **Context Menu** appears (**Figure 4.145**).
4. *Select* the **Edit Definition** option in the menu, as shown in Figure 4.145:

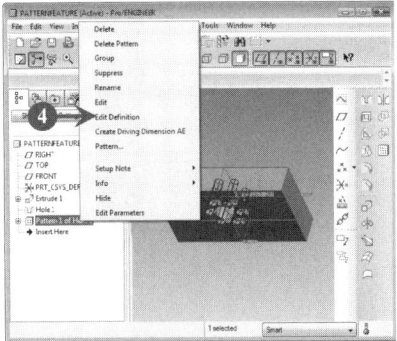

FIGURE 4.145

The dashboard corresponding to the selected feature appears. In our case, the **Pattern** dashboard appears, as shown in **Figure 4.146**.

5. *Replace* the number 6 in the **Pattern** dashboard (Figure 4.146) with the number 8 to increase the number of holes in the pattern.
6. Next, *change* the angle between the holes from 60° to 45°.
7. *Click* the **Build Feature** (☑) button, as shown in Figure 4.146:

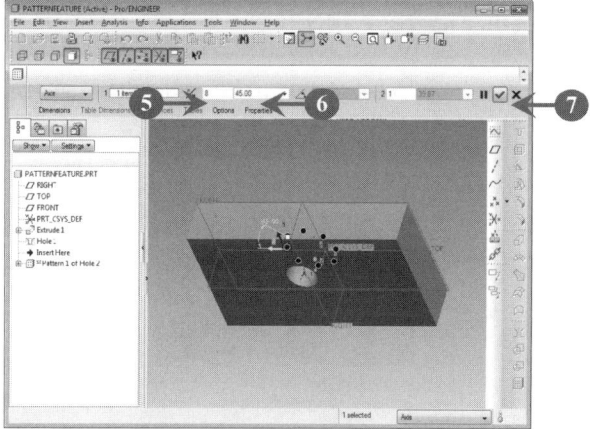

FIGURE 4.146

The modifications are reflected (in red on your screen) in the **Drawing Area**, as shown in **Figure 4.147**:

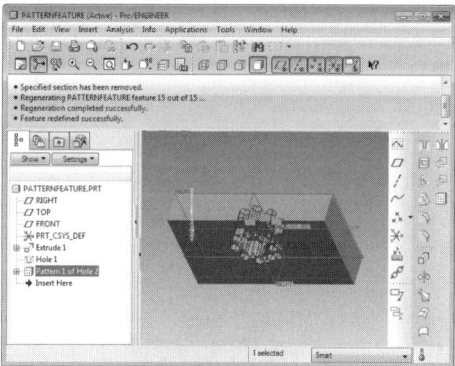

FIGURE 4.147

8. *Save* the file.

Here, we have discussed how to modify features in two ways: dimensions and definitions. With this, we complete modifying features.

Next, we study the process to delete features.

4.15 DELETING A FEATURE

During model creation, it may happen that you need to delete one or more features to get the desired model. To delete a feature from a component or a model, let's follow the steps given here:

1. *Start* the **Part** mode and *open* the HOLEFEATURE.prt file (Figure 4.42).
2. *Select* the feature to be deleted in the **Model Tree**. In our case, we have selected the feature named **Hole 1**. The selected feature is highlighted in red, as shown in **Figure 4.148**:

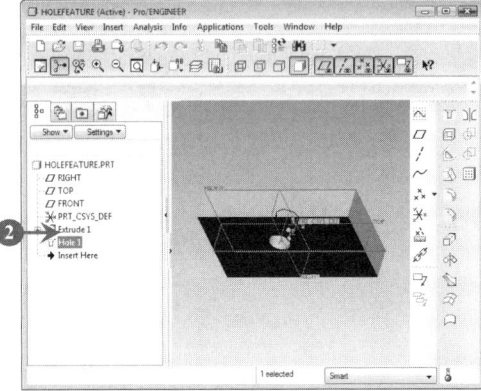

FIGURE 4.148

3. *Right-click* the selected feature (Figure 4.148). A **Context Menu** appears (**Figure 4.149**).
4. *Select* the **Delete** option, as shown in Figure 4.149:

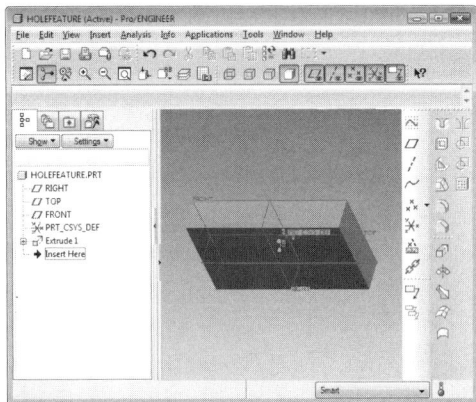

FIGURE 4.149

The **Delete dialog box** appears to confirm the deletion (**Figure 4.150**).

5. *Click* the **OK** button to delete the selected feature, as shown in Figure 4.150:

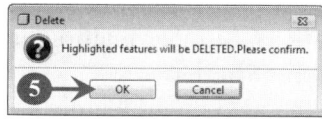

FIGURE 4.150

6. The selected feature is deleted, as shown in **Figure 4.151**:

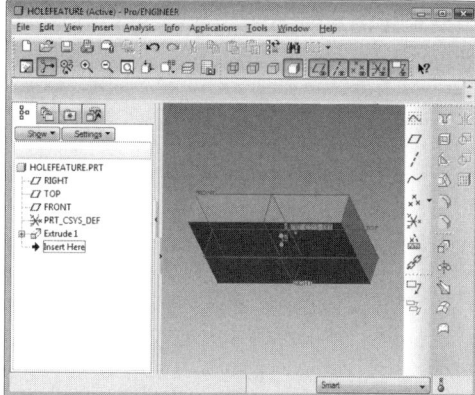

FIGURE 4.151

Note: Click the **Cancel** button to abort the process.

7. *Save* the file.

This is the process to delete a feature. With this, we come to the end of this chapter. However, before closing this chapter let's summarize the topics we have covered.

SUMMARY

In this chapter, you have learned about:

- The process to start the **Part** mode
- **Datums** and **Datum** plane creation

- The creation of a **Base Feature** and a **Protrusion**
- The process to create a **Hole** in the base feature
- The process to create a **Cut** in a feature
- The process to **Round the Edges** of a feature
- The process to create a **Chamfer** feature on a base feature
- The process to create a **Shell**
- The process to create a **Revolved** feature
- The process to create a **Sweep**
- The process to **Blend** two features
- The process to create a **Rib** to provide support to other features
- The processes to **Modify** a feature
- The process to **Delete** a feature

Chapter 5

EXPLORING PRO/ENGINEER WILDFIRE 4.0 ASSEMBLY MODE

In This Section

◊ Understanding Assembly Approaches
◊ Integrating Components in an Assembly
◊ Modifying Assembly Components
◊ Creating an Exploded View of an Assembly

The **Assembly** mode is a state in Pro/ENGINEER that a product model must pass through to complete the design process. The **Assembly** mode allows a user to assemble Part files in an Assembly file. These Part files are created either within the **Assembly** mode itself or within the **Part** mode. The **Part** mode is discussed in Chapter 4, "Exploring Pro/ENGINEER Wildfire 4.0 Part Mode". The **Assembly** mode uses either the **Top-down** or the **Bottom-up** approach to assemble components within an Assembly file. In the **Top-down** approach, components of a model are created within the **Assembly** mode. On the other hand, the **Bottom-up** approach uses existing files created within the **Part** mode in the Assembly file. The **Top-down** and **Bottom-up** approaches are discussed in detail later in this chapter.

In this chapter, you learn about the fundamentals of the **Assembly** mode in Pro/ENGINEER Wildfire 4.0. This chapter discusses the approaches used to integrate Part files in the **Assembly** mode, and how to position one Part file in relation to another in an Assembly file by using various assembly constraints to help position components within an assembly. In addition, you learn how to integrate multiple Part files in an Assembly file and modify the appearance of the Assembly file. Finally, you learn how to create an exploded view, where all the components assembled inside an Assembly file are moved from their original location so that they can be easily viewed. An exploded view helps you to view all the components placed inside an Assembly file.

You should know about the approaches supported by Pro/ENGINEER Wildfire 4.0 to assemble Part files in an Assembly file before you start assembling them.

Now, let's begin with a discussion on the approaches available in Pro/ENGINEER Wildfire 4.0 to integrate the components with the **Assembly** mode.

5.1 UNDERSTANDING ASSEMBLY APPROACHES

You learned how to create Part files in the **Part** mode in Chapter 4, "Exploring Pro/ENGINEER Wildfire 4.0 Part Mode". Part files are often created in the **Part** mode where the entities contained in them can be viewed easily to help users in designing features such as hole, cut, and shell of a Part file. However, you can also create Part files within the **Assembly** mode. Part files are assembled in the **Assembly** mode on the basis of the design mode used to create these files. Pro/ENGINEER Wildfire 4.0 provides the following approaches to integrate the components within the **Assembly** mode:

- Top-down approach
- Bottom-up approach

Let's discuss each of these approaches.

Top-Down Approach

The **Top-down** approach is a method of integrating components, such as Part files created within the **Assembly** mode. These Part files are assembled by using assembly constraints, which will be discussed later in this chapter. You must know how to create a Part file in the **Assembly** mode before you start assembling Part files.

Perform the following steps to create a Part file in the **Assembly** mode:

1. *Select* the **File > New** option from the **Menu Bar**, as shown in **Figure 5.1**:

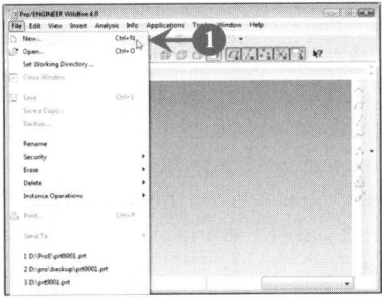

FIGURE 5.1

The **New dialog box** appears and displays the various modes available in Pro/ENGINEER Wildfire 4.0 to create a model, as shown in **Figure 5.2**:

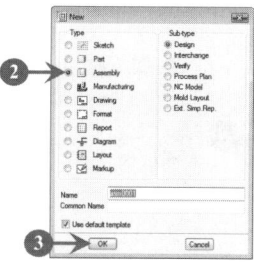

FIGURE 5.2

2. *Select* the **Assembly** radio button from the **Type** area in the **New dialog box** (Figure 5.2). The **Design** radio button is selected by default from the **Sub-type** area in the **New dialog box**. A default name (asm0001) of the Assembly file is also shown in the **Name** text box (Figure 5.2).

3. *Click* the **OK** button in the **New dialog box** (Figure 5.2). An Assembly file containing three default datum planes: ASM_RIGHT, ASM_TOP, and ASM_FRONT is created, as shown in **Figure 5.3**:

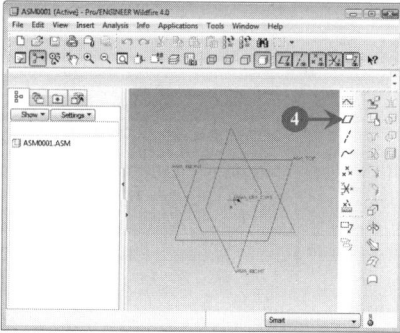

FIGURE 5.3

4. *Click* the **Create** (🔲) button from the **Tool chest** located at the right side of the **Drawing Area** (Figure 5.3). The **Component Create dialog box** appears and displays various components that you can use in the Assembly mode. You can specify a name of the Part file in the **Name** text box. In our case, the Part file is PRT0004, as shown in **Figure 5.4**:

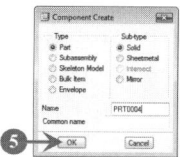

FIGURE 5.4

5. *Click* the **OK** button in the **Component Create** dialog box (Figure 5.4). The **Creation Options dialog box** appears and displays various options for creating a Part file, as shown in **Figure 5.5**:

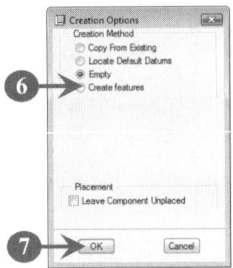

FIGURE 5.5

6. *Select* the **Empty** radio button in the **Creation Options dialog box** (Figure 5.5) to create a Part file with no initial geometry.

7. Next, *click* the **OK** button in the **Creation Options dialog box** (Figure 5.5). The Part file (PRT0004.PRT) is created in the Assembly file named ASM0001.ASM, as shown in **Figure 5.6**:

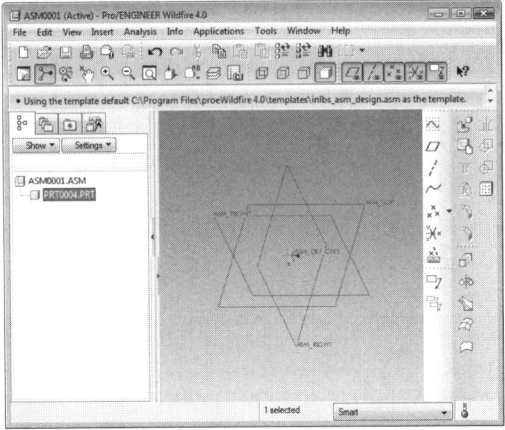

FIGURE 5.6

The PRT0004.PRT Part file displays three default datum planes, which are required to create any base feature.

8. Create a base feature using the **Extrude** feature (**Figure 5.7**) as discussed in the Chapter 4, "Exploring the Pro/ENGINEER Wildfire 4.0 Part Mode", under the section "Creating an Extruded Feature":

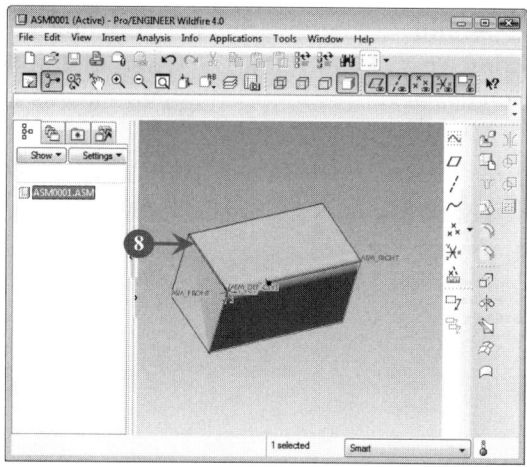

FIGURE 5.7

Now, let's discuss the **Bottom-up** approach.

Bottom-Up Approach

The **Bottom-up** approach is a method of assembling existing Part files created within the Part mode in an Assembly file. These Part files are also integrated within the Assembly file through assembly constraints. You must know how to insert an existing Part file in an Assembly file before you start assembling them.

Perform the following steps to insert an existing Part file in an Assembly file:

1. *Create* an Assembly file (.asm) by following the steps given in the preceding section "Top-Down Approach." An Assembly file containing three default datum planes is created, as shown in **Figure 5.8**:

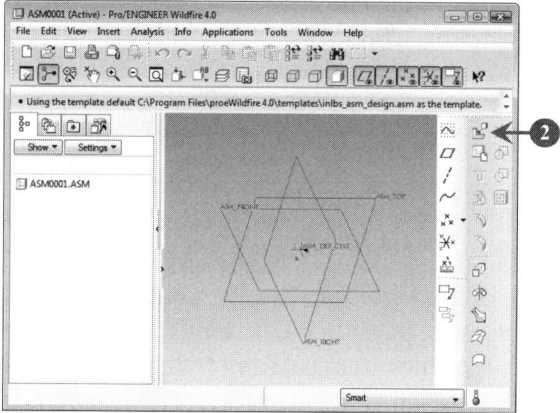

FIGURE 5.8

2. *Click* the **Assemble** (image) button from the **Tool chest** located at the right side of the **Drawing Area** (Figure 5.8). The **Open dialog box** appears, where you can browse the local system for existing Part files, as shown in **Figure 5.9**:

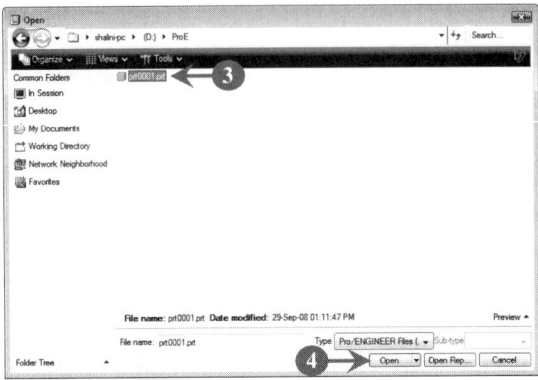

FIGURE 5.9

3. *Select* the Part file (prt0001.prt) in the **Open dialog box** (Figure 5.9).
4. Next, *click* the **Open** button in the **Open dialog box** (Figure 5.9). The Part file (prt0001.prt) is inserted in the Assembly file named ASM0001.ASM, as shown in **Figure 5.10**:

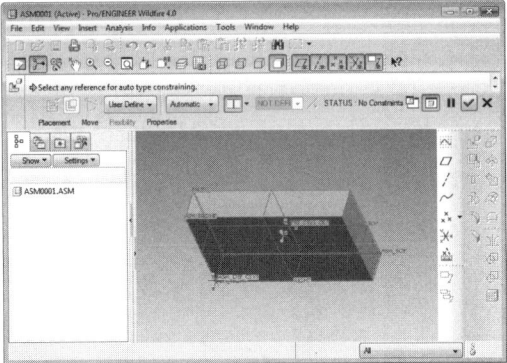

FIGURE 5.10

Part files placed in an Assembly file by using the **Assemble** button should be placed in relation to the other Part files in the Assembly file; this helps in assembling these Part files in an Assembly file. Pro/ENGINEER Wildfire 4.0 provides assembly constraints to integrate assembly components by placing these components in relation to one another in an Assembly file.

Next, let's discuss how to integrate components in an Assembly file.

5.2 INTEGRATING COMPONENTS IN AN ASSEMBLY

As discussed earlier, Pro/ENGINEER Wildfire 4.0 provides the **Top-down** and **Bottom-up** approaches for integrating or assembling Part files within an Assembly file. In addition to these approaches, there is another paradigm for integrating components within an assembly. The paradigm is based on the implementation of assembling components by using assembly constraints. If the components are assembled by using assembly constraints, the assembly created is called a parametric assembly. However, if the components are packaged and are not assembled by using the assembly constraints, the assembly created is called a nonparametric assembly.

Pro/ENGINEER Wildfire 4.0 provides the following methods to create parametric and nonparametric assemblies:

- Using Assembly constraints
- Packaging components

Let's discuss each method.

Using Assembly Constraints

After placing Part files in an Assembly file, it is important to specify how you want to arrange these Part files in relation to one another. To define the position of one Part file in relation to the other, you must use assembly constraints available in the **Constraint-type** drop-down list of the **Component Placement** dashboard, as shown in **Figure 5.11**:

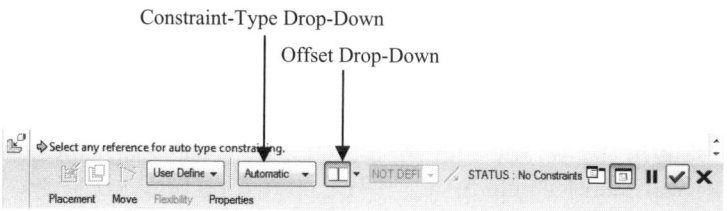

FIGURE 5.11

The **Component Placement** dashboard is located below the **Tool chest** at the top of the **Drawing Area**. It appears automatically when a Part file is placed in an Assembly file. Pro/ENGINEER Wildfire 4.0 provides various assembly constraints in the **Constraint-type** drop-down list of the **Component Placement** dashboard, as shown in **Figure 5.12**:

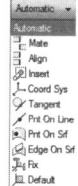

FIGURE 5.12

Figure 5.12 shows the following assembly constraints available in the **Constraint-type** drop-down list of the **Component Placement** dashboard:

■ **Automatic:** Determines automatically the appropriate constraint to apply on the selected reference. Suppose you want to align two Part files and have selected the axes of both files. In this case, Pro/ENGINEER Wildfire 4.0 automatically determines that you want to apply the **Align** constraint and applies the constraint to both Part files. The **Automatic** constraint is the default constraint in the **Component Placement** dashboard.

• **Mate:** Allows a user to position two selected references (planes or surfaces) from each Part file in such a way that they face each other. The **Mate** constraint can position the two references as coincident, orient, or offset with the help of the **Automatic** drop-down list available in the **Component Placement** dashboard (Figure 5.11). The **Mate** constraint is used in the following three combinations with the following **Offset** drop-down list options.

◊ **Mate Offset:** Allows the user to keep the two selected planes from each Part file at some distance from each other. The value of the **Offset** option determines the distance between the two planes of the Part files.

◊ **Mate Coincident:** Allows the user to position the two selected faces coplanar to each other. Coplanar faces are those faces that lie in the same plane.

◊ **Mate Oriented:** Allows the user to position the two selected planes or faces of the Part files and simultaneously orient the other planes or faces to face in the same direction.

• **Align:** Allows the user to align the two selected planes, axes, or points from each Part file to face in the same direction. An **Align** constraint can make two selected references coincident, orient, or offset with the help of the **Offset** drop-down list available in the **Component Placement** dashboard (Figure 5.11). The **Align** constraint is used in the following three combinations with the following **Offset** drop-down list options:

◊ **Align Offset:** Allows the user to align the two selected planes or faces of the Part files at some distance between them. The **Offset** value determines the distance between the two references.

◊ **Align Coincident:** Allows the user to align the two selected references in such a way that they lie in the same plane. Users can use the **Align Coincident** constraint to make the two planes coplanar.

◊ **Align Oriented:** Allows the user to align the two selected references of the Part files while orienting these references to face in the same direction.

- **Insert:** Allows the user to assemble the revolved component (a component with curved edges), such as a rod, into a hole. An **Insert** constraint makes the axes of the inserted surfaces coaxial.
- **Coord Sys:** Allows the user to align the coordinate system of a Part file with the coordinate system available in the Assembly file to place the Part file firmly in the Assembly file.
  - **Tangent:** Allows the user to place a Part file tangentially to the other Part file. A **Tangent** constraint also places the Part files in the same plane, thereby making them coplanar.
  - **Pnt On Line:** Allows the user to align the selected point on the first component (Part file) with the selected edge, axis, and datum curve on the second component (Part file).
  - **Pnt On Srf:** Allows the user to align the selected point on the first component (Part file) with the selected surface on the second component (Part file).
  - **Edge On Srf:** Allows the user to align the selected edge of the first component (Part file) with the selected surface on the second component (Part file).
  - **Default:** Allows the user to align the default coordinate system of a Part file with the default coordinate system of the assembly. This also aligns the datum planes of the Part file with the datum planes of the assembly.
- **Fix:** Allows the user to fix the current position of a moved component (Part file).

Now, let's learn how to integrate the components by using assembly constraints within an assembly file. To do so, perform the following steps:

1. *Select* the **File > New** option from the **Menu Bar**, as shown in **Figure 5.13**:

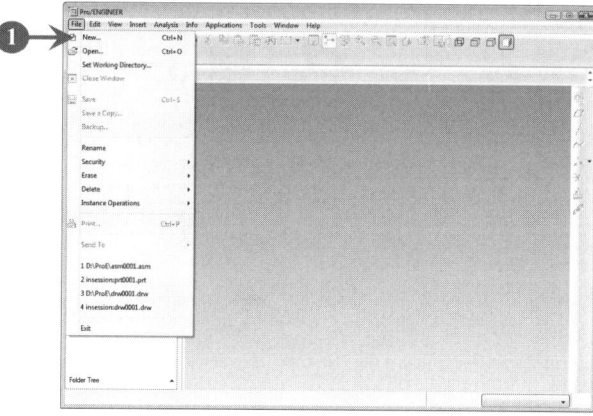

FIGURE 5.13

The **New dialog box** appears and displays the various modes available in Pro/ENGINEER Wildfire 4.0 to create a model, as shown in **Figure 5.14**:.

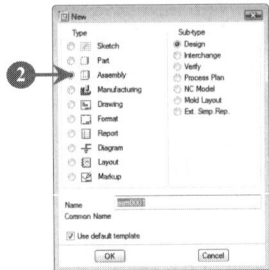

FIGURE 5.14

2. *Select* the **Assembly** radio button from the **Type** area in the **New dialog box** (Figure 5.14). The default name of the Assembly file (asm0001) is shown in the **Name** text box.

3. *Specify* the name of the Assembly file in the **Name** text box as ASSEM-BLYDEMO and *click* the **OK** button, as shown in **Figure 5.15**:

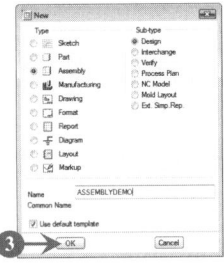

FIGURE 5.15

An Assembly file containing three default datum planes is created in the current session of Pro/ENGINEER Wildfire 4.0, as shown in **Figure 5.16**:

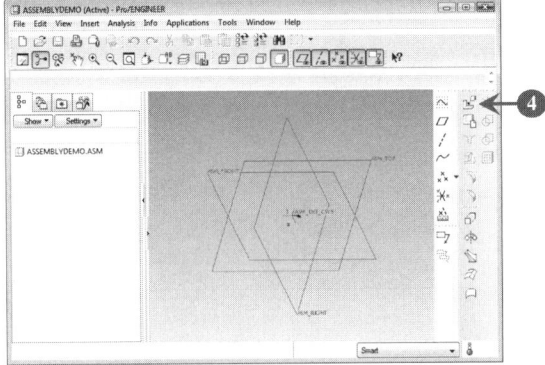

FIGURE 5.16

4. *Click* the **Assemble** ( ) button from the **Tool chest** located at the right side of the **Drawing Area** (Figure 5.16). The **Open dialog box** appears. Here, you can select the Part files that you need to assemble, as shown in **Figure 5.17**:

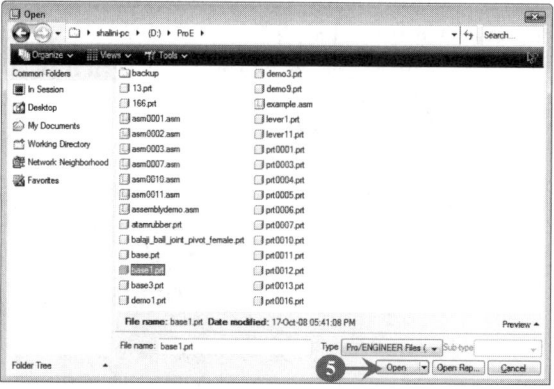

FIGURE 5.17

5. *Select* the base1.prt component in the **Open dialog box** and *click* the **Open** button (Figure 5.17).

Note: The base1.prt file already exists.

The base1.prt file is inserted into the ASSEMBLYDEMO.ASM Assembly file. The **Component Placement** dashboard appears and prompts you to select any reference for the **Automatic** constraint, as shown in **Figure 5.18**:

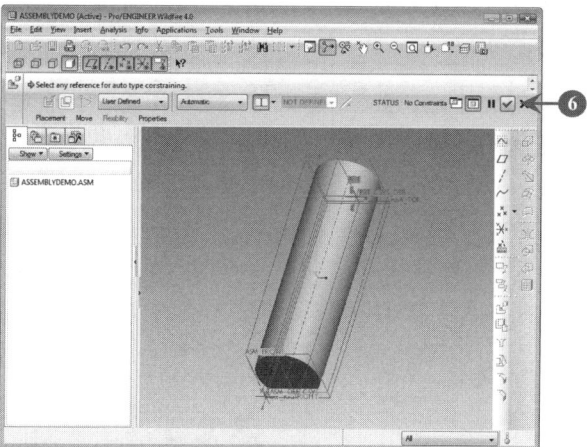

FIGURE 5.18

The **Automatic** constraint is the default constraint in the **Component Placement** dashboard, which automatically determines the appropriate constraint to apply on the selected reference.

6. *Accept* the default constraint (**Automatic**) from the **Constraint-type** drop-down list in the **Component Placement** dashboard by *clicking* the **Build feature** button (Figure 5.18). The BASE1.PRT component is successfully inserted in the Assembly file, as shown in **Figure 5.19**:

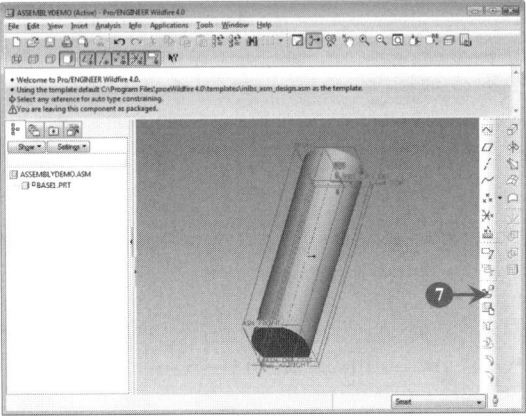

FIGURE 5.19

7. Now, let's insert a second component to the Assembly file. For this, *click* the **Assemble** (⬛) button from the **Tool chest** located at the right side of the **Drawing Area**.

The **Open dialog box** appears (**Figure 5.20**).

8. *Select* the Part file (base.prt) (Figure 5.20).
9. *Click* the **Open** button, as shown in Figure 5.20:

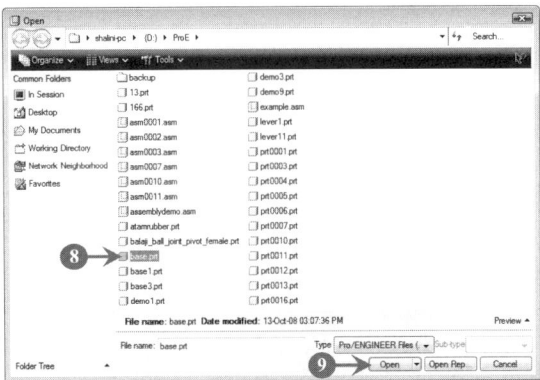

FIGURE 5.20

The second component is inserted in the Assembly file and the **Component Placement** dashboard appears. The constraints available in the **Component Placement** dashboard are used to specify how the components (base1.prt and base.prt in our case) are placed in relation to one another in the Assembly file.

10. *Select* the **Insert** option from the **Constraint-type** drop-down list in the **Component Placement** dashboard, as shown in **Figure 5.21**:

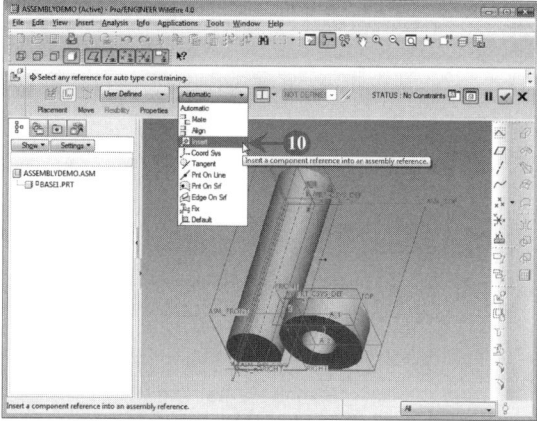

FIGURE 5.21

The **Insert** constraint is applied on the BASE.PRT component, and a prompt appears in the message area to let you select a surface on the BASE1.PRT component so that you can insert the BASE.PRT component in it (Figure 5.21).

11. *Select* the surface on the BASE1.PRT component to apply the **Insert** constraint, as shown in **Figure 5.22**:

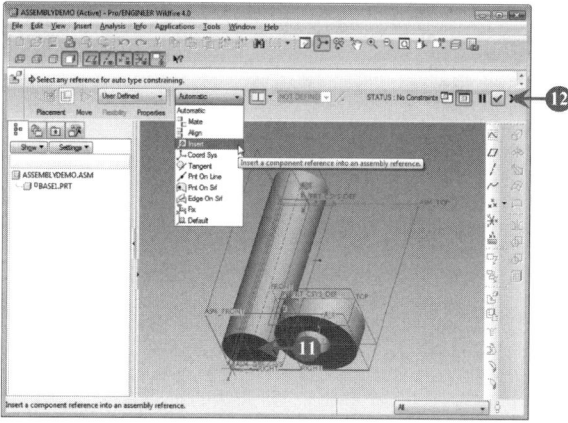

FIGURE 5.22

12. *Click* the **Build feature** (☑) button to assemble both the components (BASE1.PRT and BASE.PRT) by using the **Insert** constraint (Figure 5.22).

The assembled components are shown in **Figure 5.23**:

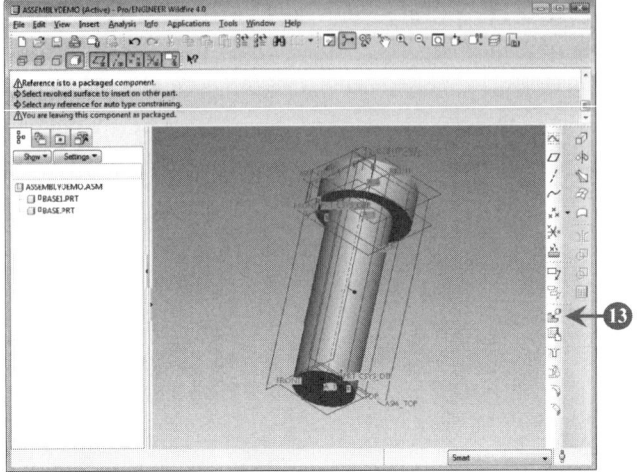

FIGURE 5.23

13. Now, let's insert the third component (BASE.PRT) in the ASSEMBLYDEMO .ASM Assembly file. To do this, *click* the **Assemble** (🖼) button (Figure 5.23).

The **Open dialog box** appears (**Figure 5.24**).

14. *Select* the base.prt file (Figure 5.24).
15. *Click* the **Open** button, as shown in Figure 5.24:

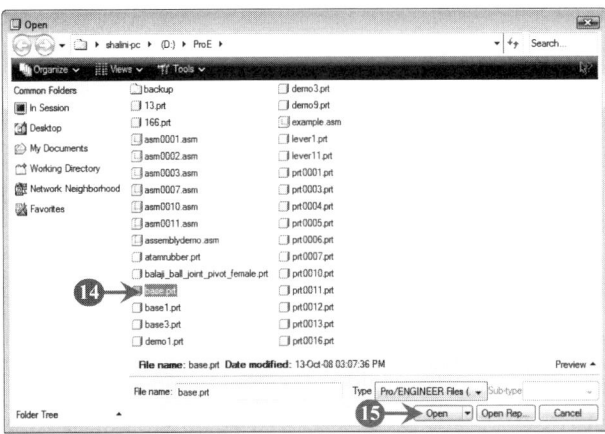

FIGURE 5.24

The BASE.PRT component is inserted in the Assembly file and the **Component Placement** dashboard appears, as shown in **Figure 5.25**:

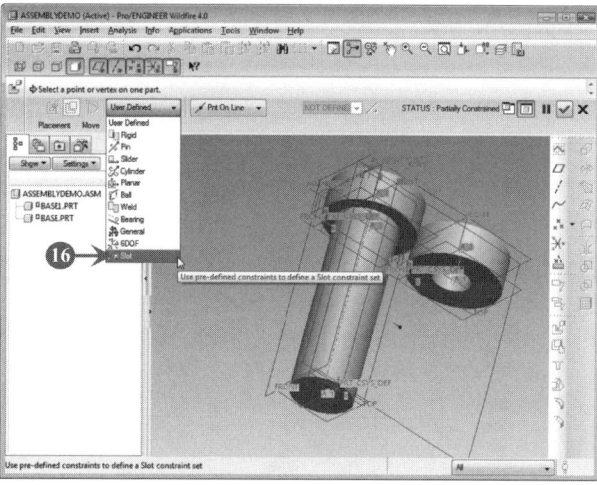

FIGURE 5.25

16. *Select* a constraint from the Automatic **Constraint-type** drop-down list. In our case, we have selected the **Pnt On Line** constraint. In addition, select the **Slot** option from the **User Defined** drop-down list (Figure 5.25). A prompt to select a point on one component appears.

17. *Select* a point on the third component (BASE.PRT) in the Assembly file. The **Pnt On Line** constraint is applied on the component. A red line is displayed to select an axis on the first component (BASE1.PRT) in the Assembly file, as shown in **Figure 5.26**:

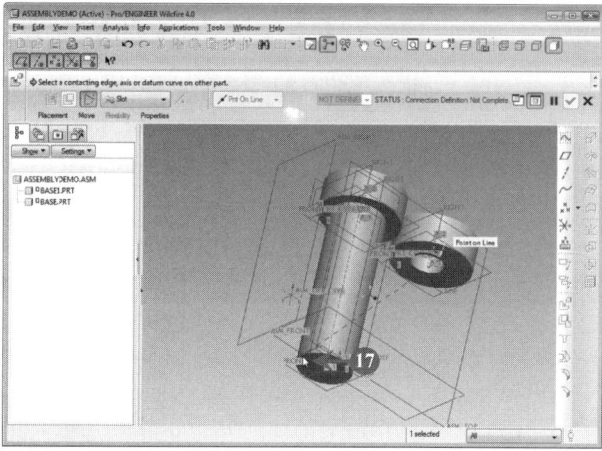

FIGURE 5.26

18. *Click* the **Build feature** (☑) button to assemble both components by using the **Pnt On Line** constraint, as shown in Figure 5.27:

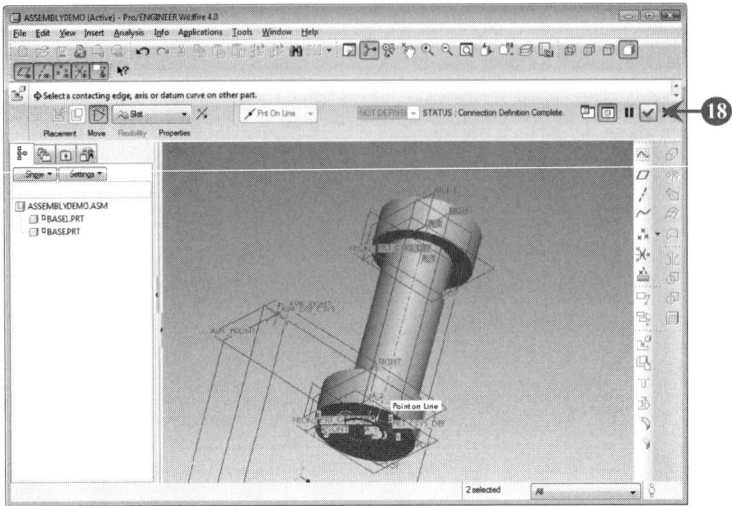

FIGURE 5.27

The selected components are assembled by using the **Pnt On Line** constraint, as shown in **Figure 5.28**:

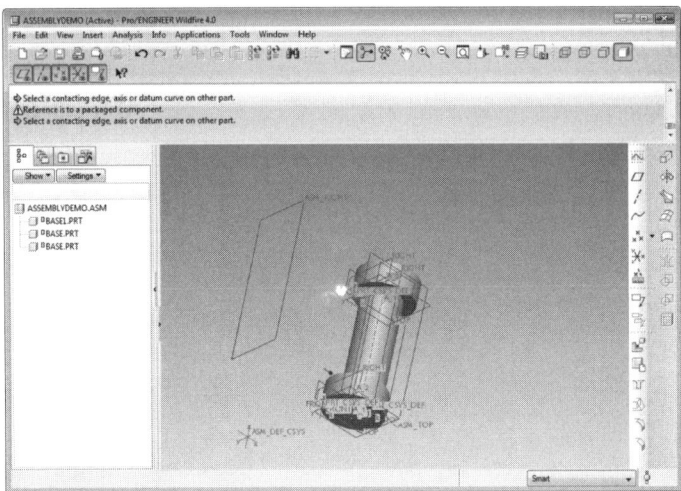

FIGURE 5.28

Pro/ENGINEER Wildfire 4.0 allows you to set the color of the components inside an assembly. Now, let's learn how to set the color of the identical components assembled in the assembly (ASSEMBLYDEMO.ASM). To learn the process to set color on assembly components, we use the ASSEMBLYDEMO.ASM file.

The following steps help you set colors on assembly components:

1.   *Select* the **View>Color and Appearance** option from the **Menu Bar**, as shown in **Figure 5.29**:

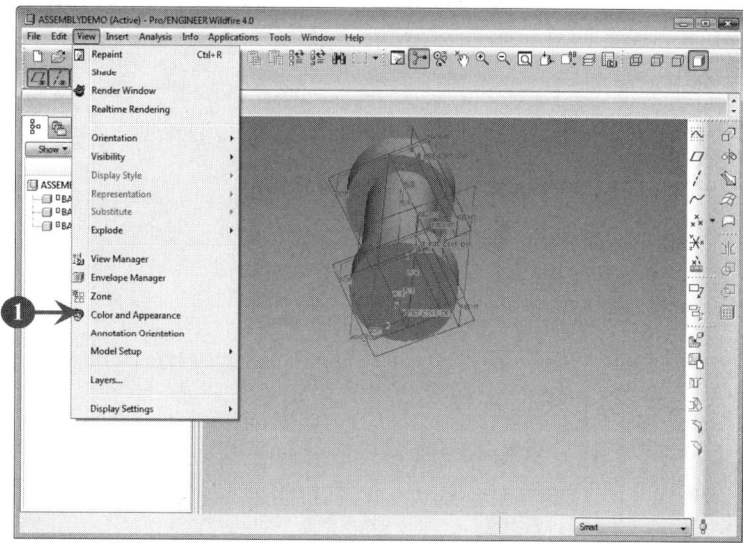

FIGURE 5.29

The **Appearance Editor window** appears and displays various options for setting the color and appearance of assembly components, as shown in **Figure 5.30**:

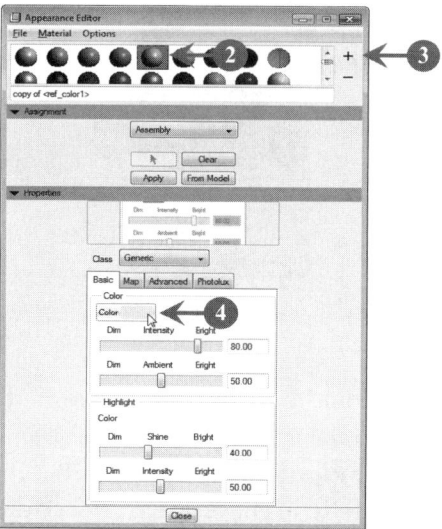

FIGURE 5.30

2. *Select* an appearance from the predefined set of appearances at the top of the **Appearance Editor window** (Figure 5.30).

3. Next, *click* the (**+**) button to create a copy of the selected color (Figure 5.30).

4. Now, *click* the **Color** button under the Basic tab of the **Properties** palette in the **Appearance Editor window** (Figure 5.30). The **Color Editor dialog box** appears, as shown in **Figure 5.31**:

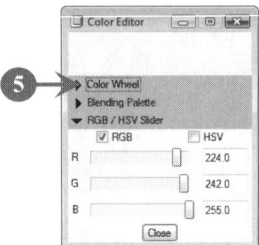

FIGURE 5.31

5. *Select* the **Color Wheel** option (Figure 5.31) to display the **Color Wheel** palette from where you can select a color for an assembly component, as shown in **Figure 5.32**:

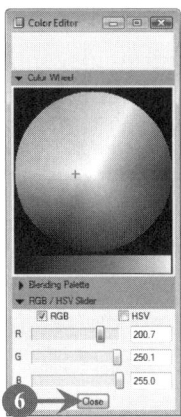

FIGURE 5.32

6. *Select* the desired color and *click* the **Close** button on the **Color Editor window** (Figure 5.32).

7. Next, *select* the **Components** option from the drop-down list available in the **Assignment** palette in the **Appearance Editor window**, as shown in Figure 5.33:

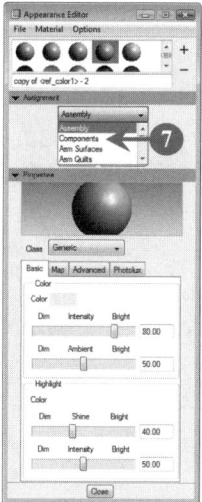

FIGURE 5.33

The **Select dialog box** appears and prompts the user to select the desired assembly component to apply the color selected from the **Color Wheel** palette, as shown in **Figure 5.34**:

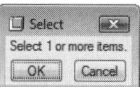

FIGURE 5.34

8. *Select* any of the BASE.PRT components from the **Drawing Area** to set the color for the component, as shown in **Figure 5.35**:

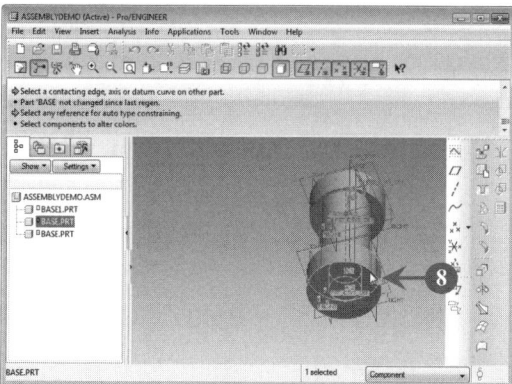

FIGURE 5.35

The selected component is highlighted with a blue outline (Figure 5.35).

9. Next, *click* the **OK** button to close the **Select dialog box**, as shown in **Figure 5.36**:

FIGURE 5.36

10. Now, to apply the color on the selected component, *click* the **Apply** button in the **Assignment** palette in the **Appearance Editor window**, as shown in **Figure 5.37**:

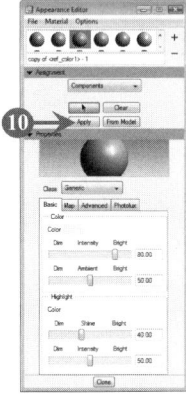

FIGURE 5.37

The specified color is applied to the selected component (BASE.PRT), as shown in **Figure 5.38**:

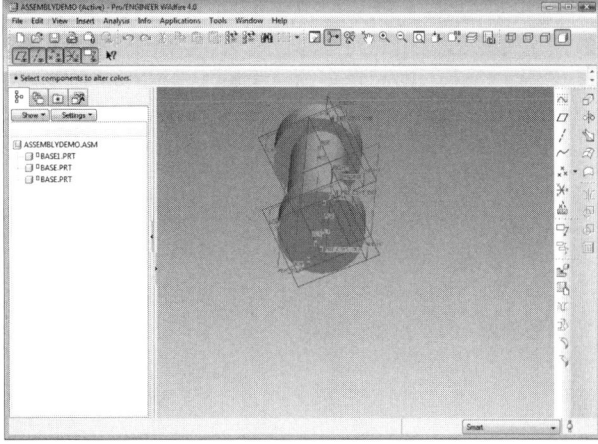

FIGURE 5.38

11. Now, you need to apply color to the other identical component (BASE.PRT). To do this, *click* the **Select** (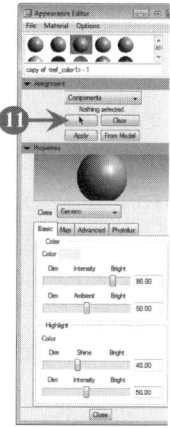) button from the **Assignment** palette in the **Appearance Editor window**, as shown in **Figure 5.39**:

FIGURE 5.39

The **Select dialog box** reappears, prompting you to select the identical component (BASE.PRT) that you want to color, as shown in **Figure 5.40**:

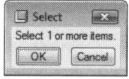

FIGURE 5.40

12. *Select* the identical BASE.PRT component from the **Drawing Area** to set the specified color for the component, as shown in **Figure 5.41**:

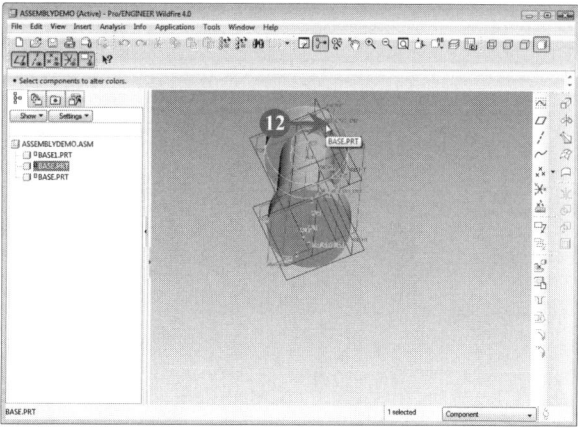

FIGURE 5.41

13. Next, *click* the **OK** button in the **Select dialog box**, as shown in **Figure 5.42**:

FIGURE 5.42

This closes the **Select dialog box**. Now, to complete the process of applying the color to the selected component, move to the next step.

14. *Click* the **Apply** button from the **Appearance Editor window**, as shown in **Figure 5.43**:

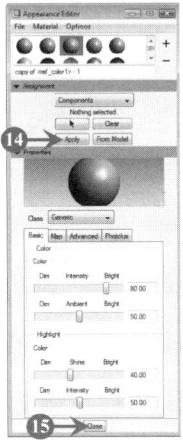

FIGURE 5.43

15. *Click* the **Close** button in the **Appearance Editor window** to close it (Figure 5.43).

The specified color is applied to the selected component, as shown in **Figure 5.44**:

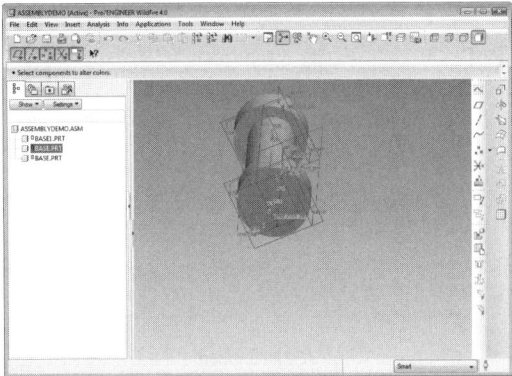

FIGURE 5.44

16. After changing the appearance of the components in the ASSEMBLYDEMO .ASM file, let's save the changes made to the file. For this, *select* the **File**>**Save** option from the **Menu Bar**, as shown in **Figure 5.45**:

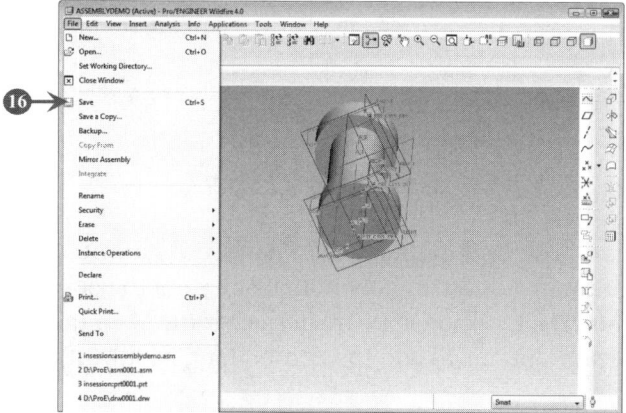

FIGURE 5.45

Next, we discuss how to package components in an assembly.

Packaging Components

Packaging is a mechanism of assembling components when their exact positions in the Assembly file are not known. In such cases, the components cannot be assembled by using assembly constraints and can only be packaged.

Perform the following steps to package a component in an Assembly file:

1. *Create* an Assembly file and *select* the **Insert**>**Component**>**Package** option from the **Menu Bar**, as shown in **Figure 5.46**:

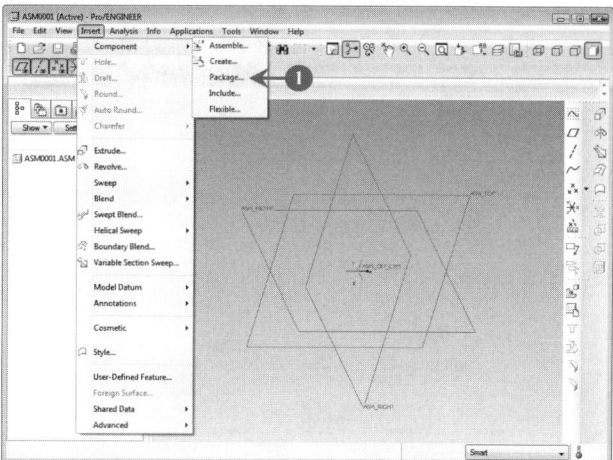

FIGURE 5.46

The **Menu Manager menu** appears and displays various options in the **PACKAGE menu**, as shown in **Figure 5.47**:

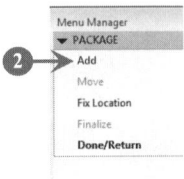

FIGURE 5.47

2. *Select* the **Add** option under the **PACKAGE menu** from the **Menu Manager menu** to add a component that you want to package (Figure 5.47). The **GET MODEL** option appears in the **Menu Manager menu** and displays various sub-options to add the component, as shown in **Figure 5.48**:

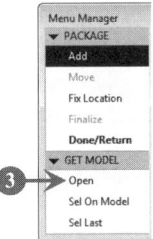

FIGURE 5.48

3. *Select* the **Open** sub-option under the **GET MODEL** option in the **Menu Manager menu** (Figure 5.48). The **Open dialog box** appears and displays the existing Part files available for packaging in the Assembly file (ASM0001.ASM in our case), as shown in **Figure 5.49**:

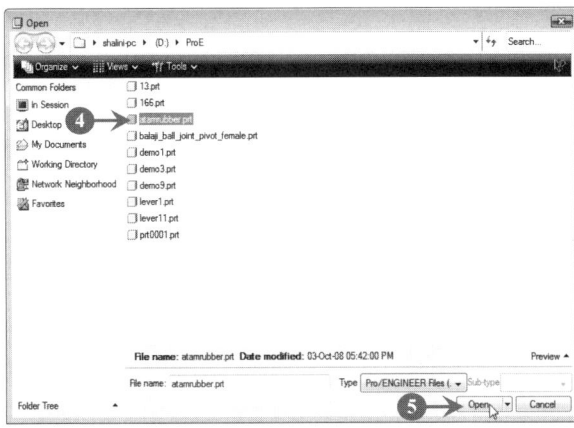

FIGURE 5.49

4. *Select* the Part file that you want to package from the **Open dialog box**. In our case, we have selected the atamrubber.prt Part file (Figure 5.49).

5. Next, click the Open button (Figure 5.49).

The selected Part file is placed in the ASM0001.ASM Assembly file, and the **Move dialog box** appears. You can use the **Move dialog box** to move a packaged component to a new location (**Figure 5.50**).

6. *Select* a radio button from the **Motion Type** area in the **Move dialog box** to specify the type of motion you want to apply to the component. By default the **Orient Mode** radio button is selected and we continue with it (Figure 5.50).

7. Now, click the **OK** button, as shown in Figure 5.50:

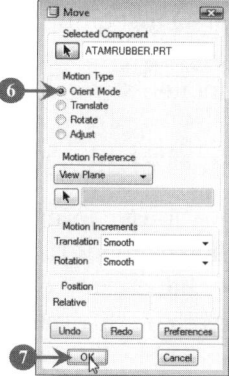

FIGURE 5.50

The selected component (ATAMRUBBER.PRT) is packaged successfully in the Assembly file, as shown in **Figure 5.51**:

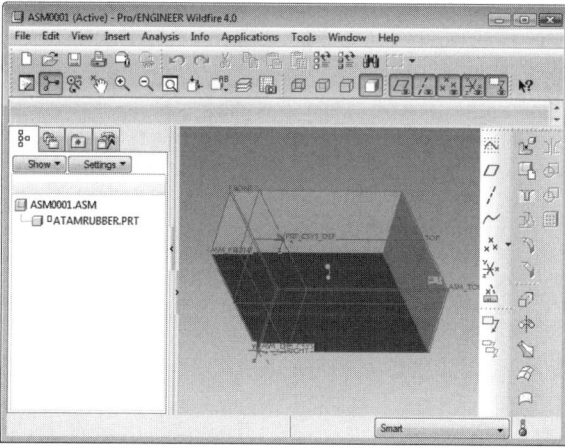

FIGURE 5.51

Now, let's discuss how to modify assembly components within an assembly.

5.3 MODIFYING ASSEMBLY COMPONENTS

Users often face a situation where they want to edit the features of assembly components in the **Assembly** mode. This may happen when the result from a Part file is not reflected in the **Assembly** mode. A user can perform this task easily in Pro/ENGINEER Wildfire 4.0 by redefining the features of the assembled components in the **Assembly** mode. However, you can also edit the assembly components in the **Part** mode by modifying the dimensions of the features of the assembled components.

Pro/ENGINEER Wildfire 4.0 allows you to perform the following manipulations on assembly components:

- Modify the dimensions of a feature of an assembly component
- Redefine the features of an assembly component

Let's discuss each of these components.

Modifying Dimensions of a Feature of an Assembly Component

Pro/ENGINEER Wildfire 4.0 allows a user to modify the dimensions of the features of an assembly component in the **Part** mode. Let's learn how to do this by using the ASSEMBLYDEMO.ASM Assembly file created earlier in this chapter.

Now, perform the following steps to modify the dimensions of a component feature in the ASSEMBLYDEMO.ASM Assembly file:

1. *Select* the **File > Open** option from the **Menu Bar** to open the Assembly file in which you want to make modifications, as shown in **Figure 5.52**:

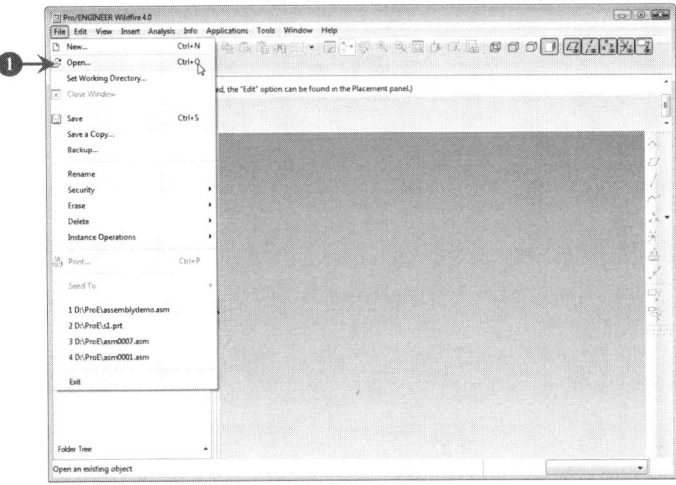

FIGURE 5.52

The **Open dialog box** appears and displays a list of Pro/ENGINEER Wildfire 4.0 files, as shown in **Figure 5.53**:

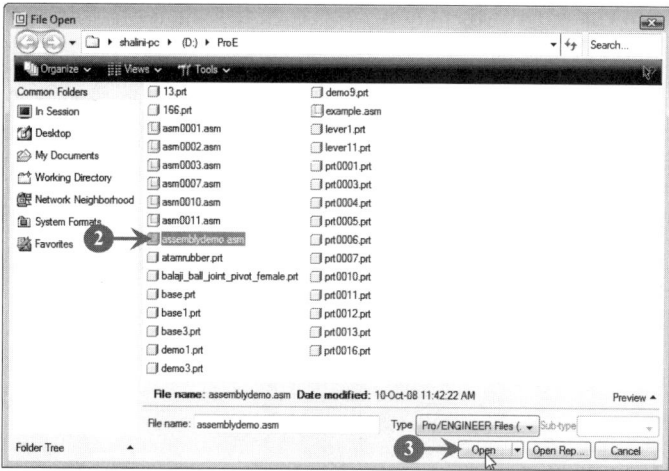

FIGURE 5.53

2. *Select* the assemblydemo.asm Assembly file (Figure 5.53).
3. *Click* the **Open** button (Figure 5.53). The selected Assembly file opens in the current session of Pro/ENGINEER Wildfire 4.0, as shown in **Figure 5.54**:

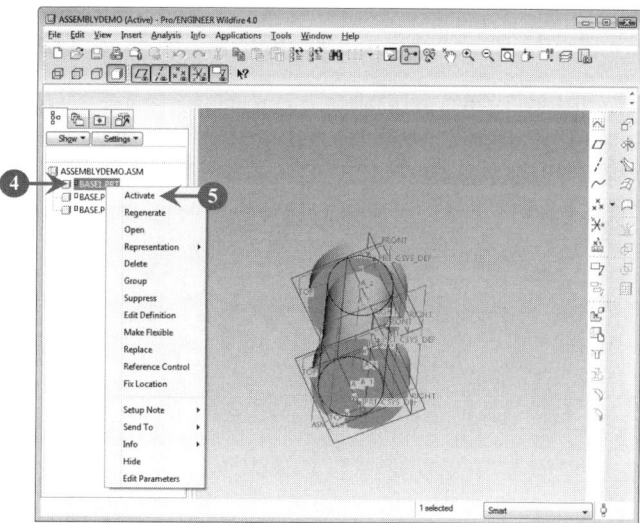

FIGURE 5.54

4. *Right-click* the desired component (BASE1.PRT) from the **Model Tree** located at the left side of the **Drawing Area**. A context menu appears (Figure 5.54).

5. *Select* the **Activate** option from the context menu (Figure 5.54). A small green button appears on the selected component in the **Model Tree** and the components other than those of the BASE1.PRT component become faded in the **Drawing Area**, as shown in **Figure 5.55**:

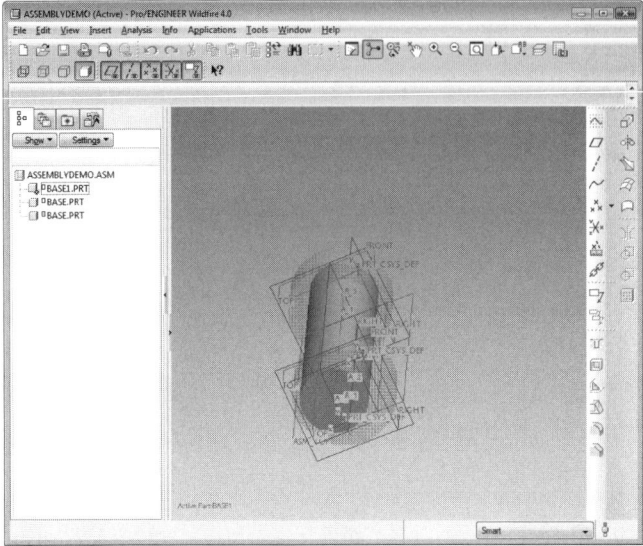

FIGURE 5.55

6. *Right-click* the selected component (BASE1.PRT) and select the **Open** option from the context menu, as shown in **Figure 5.56**:

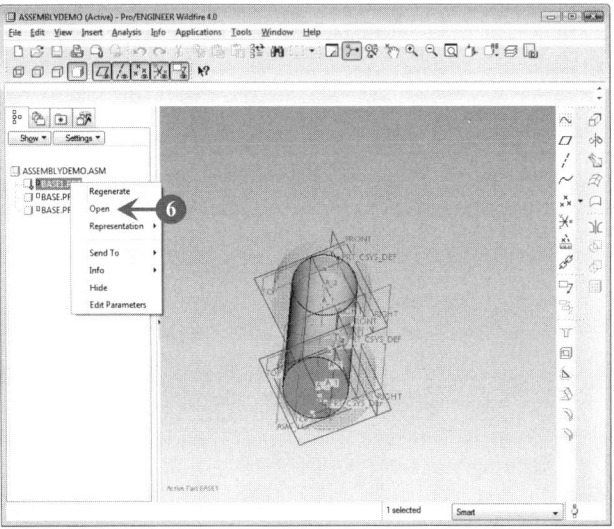

FIGURE 5.56

The selected component opens as a Part file in a separate window, as shown in **Figure 5.57**:

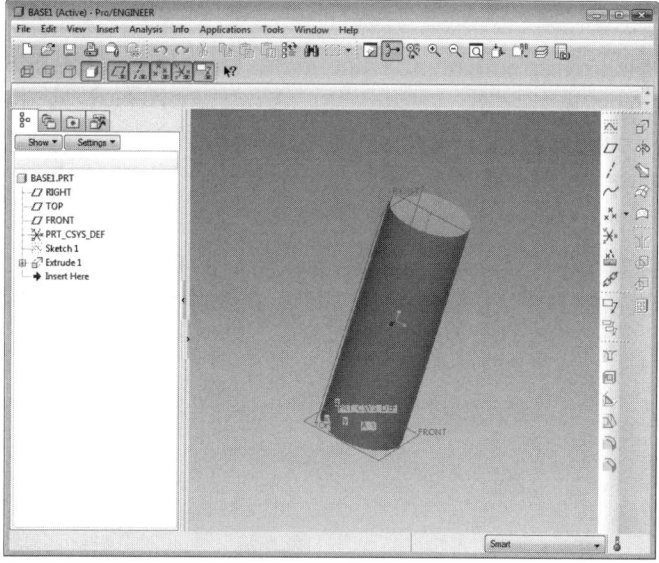

FIGURE 5.57

7. *Right-click* the desired feature (**Extrude**) in the **Model Tree** and *select* the **Edit Definition** option from the context menu, as shown in Figure 5.58:

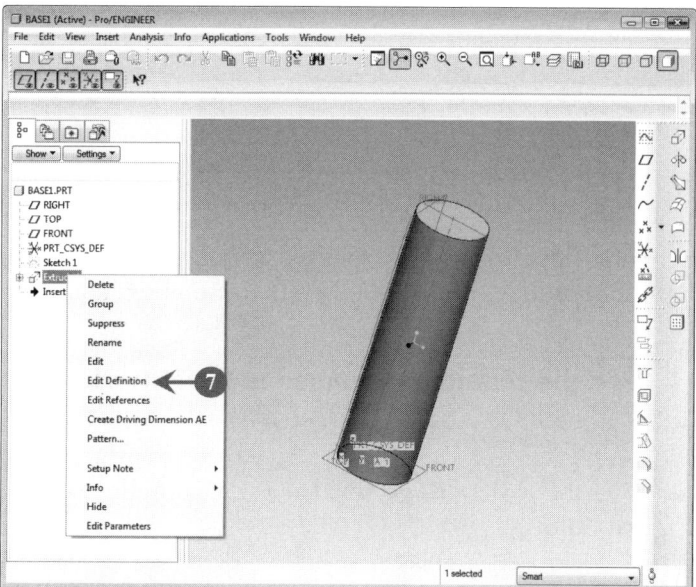

FIGURE 5.58

The **Extrude** dashboard appears (**Figure 5.59**). Here, you can modify the dimensions of the **Extrude** feature in the BASE1.PRT component by changing its existing depth value.

8. *Enter* an appropriate depth value in the **Depth** box available in the **Extrude** dashboard to modify the dimensions of the **Extrude** feature. In our case, we have entered the value 400.00 (Figure 5.59).

9. *Press* the **ENTER** key to reflect the new depth value in the **Extrude** dashboard.

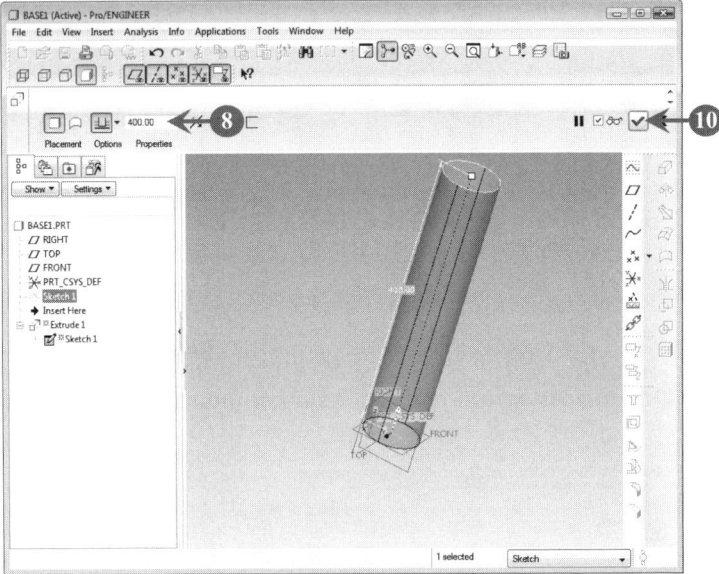

FIGURE 5.59

The depth value entered in the **Extrude** dashboard modifies the previous depth value of the **Extrude** feature applied on the assembled component.

10. *Click* the **Build feature** (☑) button in the **Extrude** dashboard to apply and save the changes made in the **Extrude** feature of the component (Figure 5.59).

11. Next, *select* the **Edit > Regenerate** option from the **Menu Bar**, as shown in **Figure 5.60**:

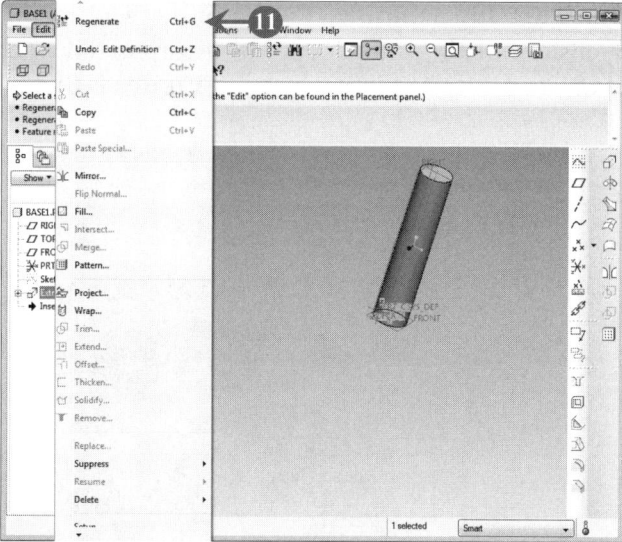

FIGURE 5.60

The component (BASE1.PRT) is regenerated (that is, the modifications made in the dimensions of the **Extrude** feature are permanently saved in memory), as shown in **Figure 5.61**:

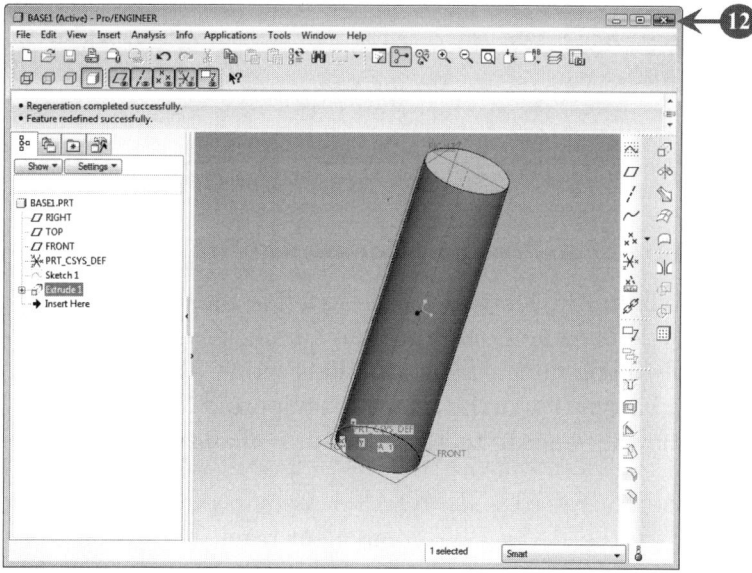

FIGURE 5.61

12. *Close* the **Part file window** (Figure 5.61). The modifications made in the dimensions of the component are incorporated in the **Assembly** file (ASSEMBLYDEMO.ASM), as shown in **Figure 5.62**:

FIGURE 5.62

Now, let's discuss how to redefine a feature of a component within an assembly.

Redefining a Feature of an Assembly Component

Pro/ENGINEER Wildfire 4.0 allows a user to redefine a feature of a component in the **Assembly** mode. Redefining here simply means editing a feature of a component while it is in the **Assembly** mode rather than in the **Part** mode. This allows a user to get the expected result in a Part file when the Part file is imported to an assembly.

Perform the following steps to redefine a feature of an assembly component:

1. *Open* the ASSEMBLYDEMO.ASM Assembly file.
2. *Right-click* the BASE.PRT component from the **Model Tree**, and *select* the **Activate** option from the context menu. A small green button appears on the selected component in the **Model Tree** and the components other than the selected component appear faded in the **Drawing Area**, as shown in **Figure 5.63**:

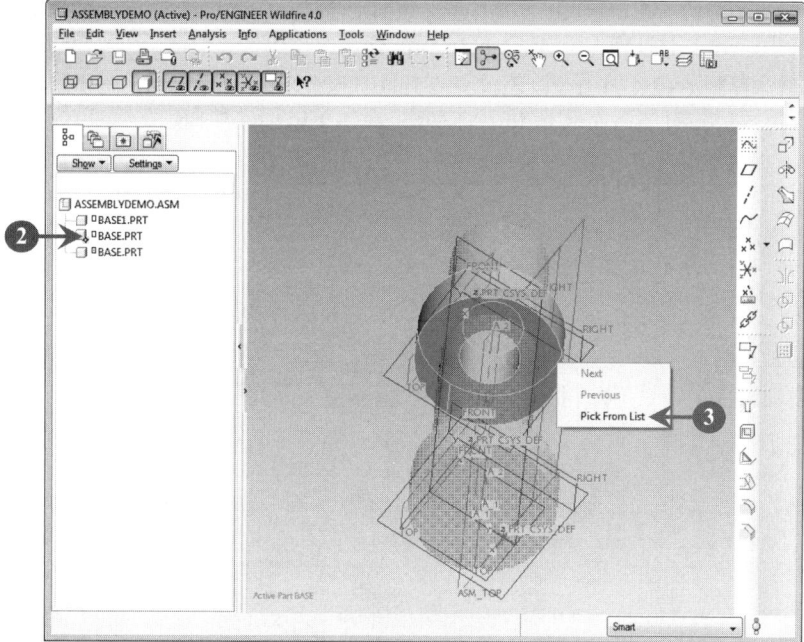

FIGURE 5.63

3. *Right-click the* BASE1.PRT component and *select* the **Pick From List** option from the context menu (Figure 5.63). The **Pick From List dialog box** appears, as shown in **Figure 5.64**:

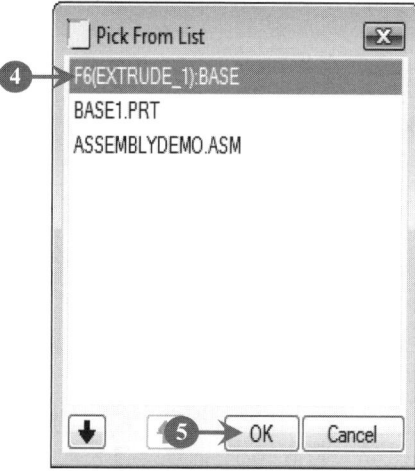

FIGURE 5.64

The **Pick From List dialog box** is used to redefine a feature of a selected component.

4. *Select* the **Extrude** feature of the selected component (Figure 5.64), F6(EXTRUDE_1): BASE in our case.

5. *Click* the **OK** button (Figure 5.64). The component whose **Extrude** feature is selected appears faded, as shown in **Figure 5.65**:

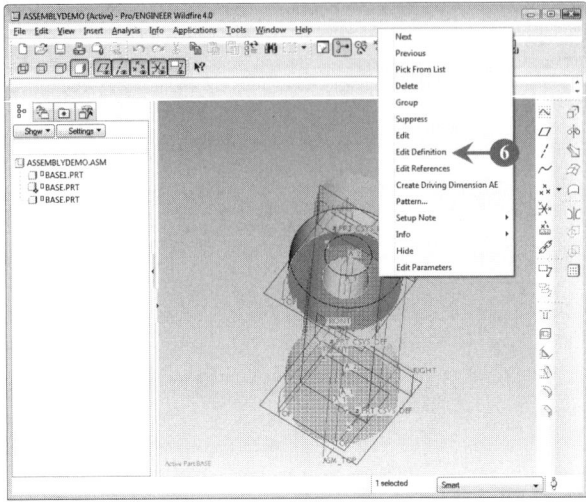

FIGURE 5.65

6. *Right-click* the **Drawing Area** and select the **Edit Definition** option from the context menu (Figure 5.65).

7. Next, *specify* a depth value as per your requirement in the **Depth box** of the **Extrude** dashboard, as shown in **Figure 5.66**:

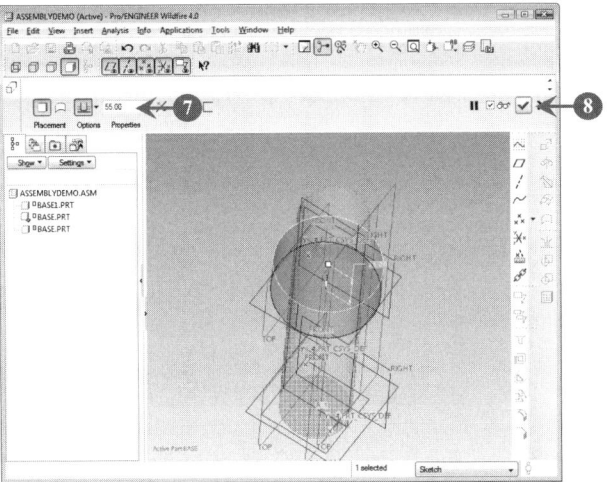

FIGURE 5.66

The depth value that you specify in the **Extrude** dashboard modifies the previous depth value of the selected component.

8. *Click* the **Build feature** ☑ button in the **Extrude** dashboard to apply and save the changes made in **Extrude** feature of the selected component (Figure 5.66). The modifications which are made in the **Extrude** feature of the BASE.PRT component are then incorporated within the Assembly file ASSEMBLYDEMO.ASM, as shown in **Figure 5.67**:

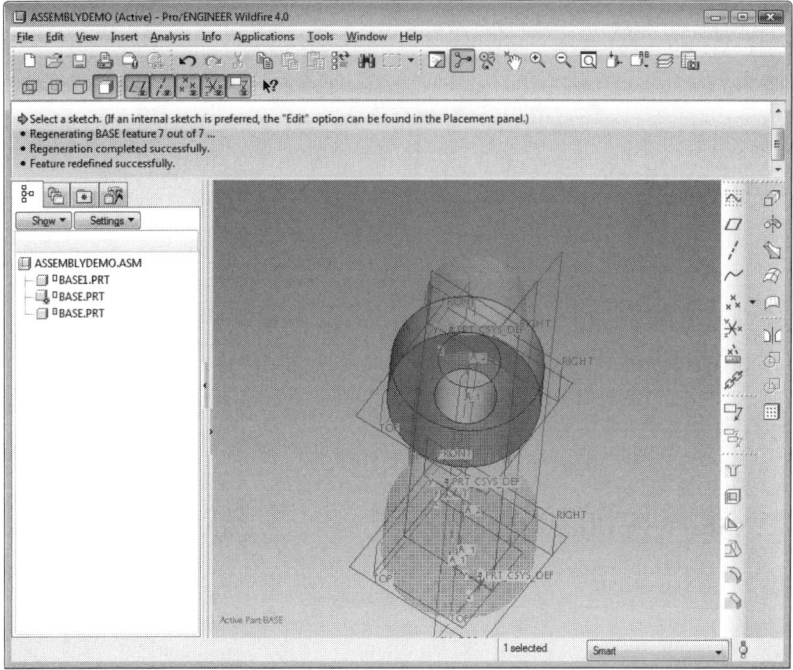

FIGURE 5.67

We learned about modifying assembly components in the previous section. Now, let's discuss how to create an exploded view of the Assembly.

5.4 CREATING AN EXPLODED VIEW OF AN ASSEMBLY

Users often find it difficult to view all the components placed inside an Assembly file, which, in turn, can lead a user to misinterpret the information intended to be conveyed. For example, a wrong value for a dimension can be entered by a user. Pro/ENGINEER Wildfire 4.0 provides a solution to such problems by creating an exploded view of an assembly. The exploded view is a state in which all the components assembled inside an

Assembly file are moved from their actual location so that they can be easily viewed. Let's take a look at the steps required to create an exploded view of the ASSEMBLYDEMO .ASM Assembly file created earlier in this chapter.

Perform the following steps to create an exploded view of the ASSEMBLYDEMO.ASM Assembly file:

1. *Select* the **View** > **Explode** > **Edit Position** option from the **Menu Bar**, as shown in **Figure 5.68**:

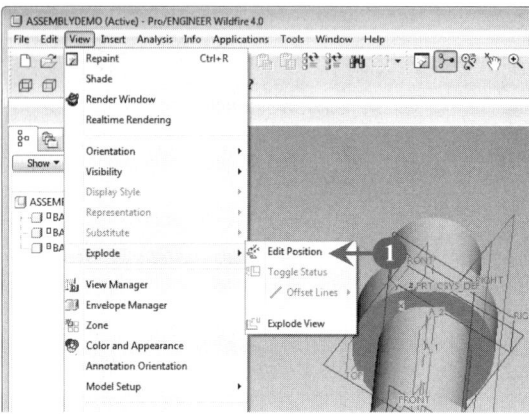

FIGURE 5.68

The **Explode Position dialog box** appears, as shown in **Figure 5.69**:

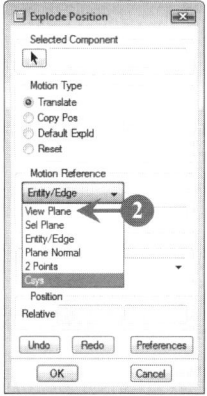

FIGURE 5.69

The **Select dialog box** also appears along with the **Explode Position dialog box**, as shown in **Figure 5.70**:

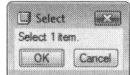

FIGURE 5.70

2. *Select* the **View Plane** option by clicking the **Entity/Edge** drop-down list in the **Motion Reference** area of the **Explode Position dialog box** (Figure 5.69).

3. Now, *select* any component; and, keeping the **mouse** button pressed, *drag* the component to the desired location. After reaching the location, release the **mouse** button, as shown in **Figure 5.71**:

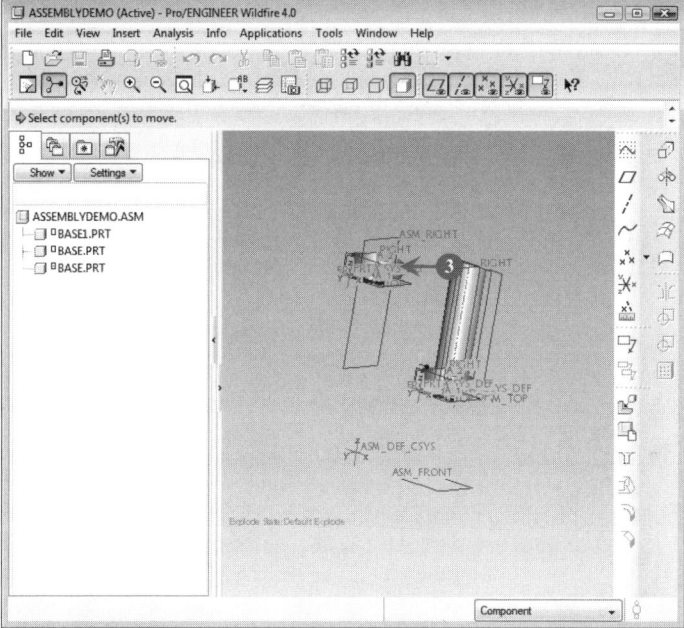

FIGURE 5.71

When you release the **mouse** button, the **Menu Manager menu** appears, as shown in **Figure 5.72**:

FIGURE 5.72

The **Menu Manager menu** helps you select another component to move. The following two options are present in the **Menu Manager menu**:

- **Select:** Allows you to select and move another component
- **Use Previous:** Allows you to select and move the previously moved component

By default the **Select** option is selected, and we continue with this option.

Repeat step 3 to move the other components in the ASSEMBLYDEMO.ASM Assembly file to display their exploded view (Figure 5.71). In our case, we have moved another component, as shown in **Figure 5.73**:

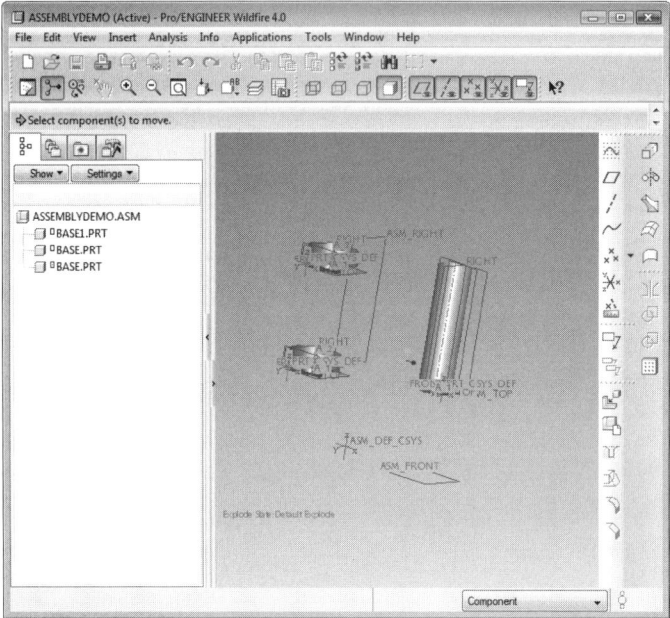

FIGURE 5.73

4. *Click* the **OK** button (Figure 5.70).
5. *Click* the **OK** button (Figure 5.69) to complete the process.

With this, we come to the end of this chapter. Before moving to the next chapter, however, let's recap the main points covered through a brief summary.

SUMMARY

In this chapter, you learned about:

- Assembly approaches, including a **Top-down** and **Bottom-up** approach to assemble components in an assembly
- Integrating assembly components with the help of assembly constraints and by packaging assembly components
- Manipulating components of an assembly by modifying the dimensions of their features and by redefining their features
- Creating an exploded view of an assembly to move the assembly components from their actual location so that they can be easily viewed in an assembly

EXPLORING PRO/ENGINEER WILDFIRE 4.0 DRAWING MODE

Chapter **6**

In This Section

◇ Starting the Drawing Mode
◇ Generating Drawing Views
◇ Modifying the Drawing Views
◇ Inducing Dimensions and Tolerance
◇ Creating Notes

Pro/ENGINEER Wildfire 4.0 allows users to create drawing views, which contain complete and detailed information about the geometry of a component (Part) to manufacture it easily. The drawing views are created to convey all necessary information about a component for proper and easy manufacturing of the component. Pro/ENGINEER Wildfire 4.0 provides the **Drawing** mode to generate the drawing views of the Parts and the Assemblies created in a Pro/ENGINEER Wildfire 4.0 session.

In this chapter, you learn the fundamentals of the **Drawing** mode in Pro/ENGINEER Wildfire 4.0. To generate the drawing view, you need to first start the **Drawing** mode. Therefore, this chapter begins with a discussion on starting the **Drawing** mode. The chapter then proceeds further to discuss the various drawing views. You also learn to modify the drawing views and induce dimensions and tolerances in the drawing views. Toward the end of the chapter, you learn about modifying dimensions and creating notes in the drawing views.

6.1 STARTING THE DRAWING MODE

The **Drawing** mode is used to generate the drawing view from the Parts and Assemblies created in a Pro/ENGINEER Wildfire 4.0 session. To generate the drawing view, first, you need to start the **Drawing** mode. Let's take a look at the steps required to start the **Drawing** mode.

Perform the following steps to start the **Drawing** mode:

1. *Select* the **File** > **New** option from the **Menu Bar**, as shown in the **Figure 6.1**:

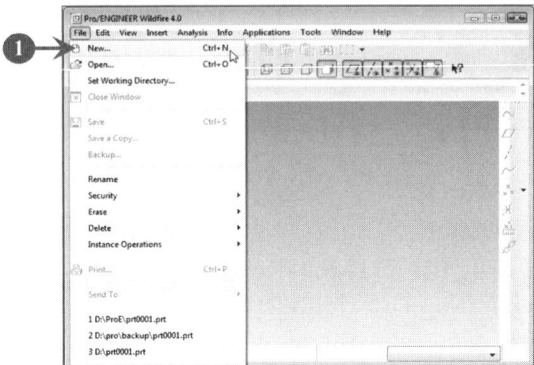

FIGURE 6.1

The **New dialog box** appears and displays the various modes available in Pro/ENGINEER Wildfire 4.0 (**Figure 6.2**).

2. *Select* the **Drawing** radio button from the **Type** area in the **New dialog box**, as shown in Figure 6.2:

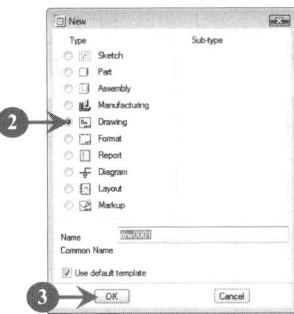

FIGURE 6.2

A default name of the Drawing file is shown in the **Name** text box. We continue with the default name. However, you can change the default name.

3. *Click* the **OK** button in the **New dialog box** (Figure 6.2). The **New Drawing dialog box** appears and displays various options to specify a template, which will be applied to a new Drawing file in a Pro/ENGINEER Wildfire 4.0 session, as shown in **Figure 6.3**:

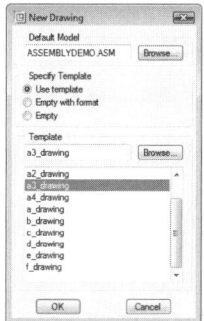

FIGURE 6.3

The **Default Model** area in the **New Drawing dialog box** allows the user to specify the name of the model for which views are to be generated.

Note: By default a model name appears in the **Default Model** area. To change the default model name, click the **Browse** button and select the model.

In our case, ASSEMBLYDEMO.ASM is displayed by default in the **Name** edit box in the **Default Model** area because it is opened in the current session of Pro/ENGINEER Wildfire 4.0. In our case, we will use the ASSEMBLYDEMO .ASM file that has already been created in Chapter 5, "Exploring Pro/ENGINEER Wildfire 4.0 Assembly Mode".

4. *Select* one of the following radio buttons from the **Specify Template** area in the **New Drawing dialog box**:

■ **Use Template:** Allows users to select any predefined drawing template for the new Drawing file (Figure 6.3). It is selected by default in the **Specify Template** area.

■ **Empty With Format:** Allows users to select a system-defined format for the new Drawing file, as shown in **Figure 6.4**:

FIGURE 6.4

- **Empty:** Allows user to create a new Drawing file with a specified size and orientation, as shown in **Figure 6.5**:

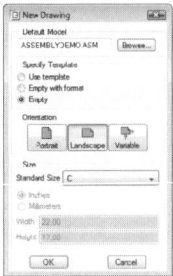

FIGURE 6.5

Once the desired option is selected from the **Specify Template** area in the **New Drawing dialog box**, the next step is to apply it. In our case, we have *selected* **Use Template** option from the **Specify Template** area.

5. *Click* the **OK** button in the **New Drawing dialog box**, as shown in **Figure 6.6**:

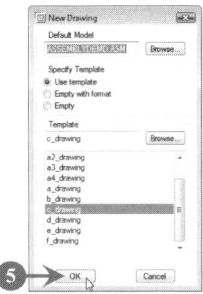

FIGURE 6.6

A Drawing file with a size and orientation specified in the predefined template selected is created, as shown in **Figure 6.7**:

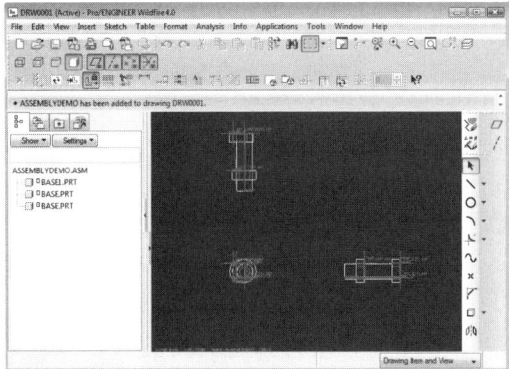

FIGURE 6.7

Now that the Drawing file has been created, you can now generate drawing views. Next, let's discuss generating drawing views in the following section.

6.2 GENERATING DRAWING VIEWS

Drawing views convey the necessary information about a component for proper and easy manufacturing of the component. There are several different types of drawing views that you can generate in a Drawing file. However, it is important to generate the **General** view before placing any other view in the Drawing file. Pro/ENGINEER Wildfire 4.0 provides the following drawing views:

- General view
- Projection view
- Detailed view
- Auxiliary view
- Revolved view

Now, let's discuss each of these drawing views.

General View

A **General** view is the first view that must be placed in the Drawing file. The **General** view, by default, is placed in its default orientation. However, you can reorient it using the default datum planes or the predefined named views. Let's take a look at the steps required to generate a **General** view.

Perform the following steps to generate a **General** view:

1. *Create* a new Drawing file named GENERALVIEW.DRW with the default model named ASSEMBLYDEMO.ASM.

2. In the Drawing file *select* the **Insert > Drawing View > General** option from the **Menu Bar**, as shown in **Figure 6.8**:

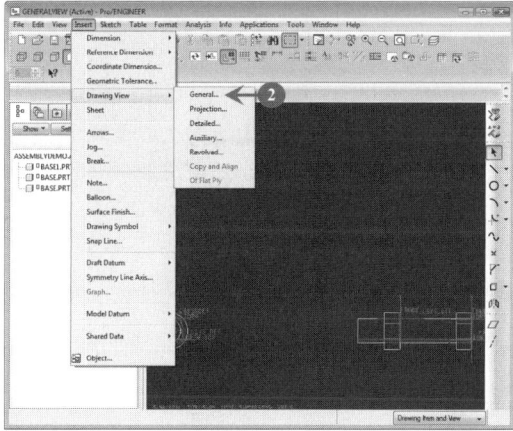

FIGURE 6.8

The **Select Combined State dialog box** appears and displays two options in the **Combined state names** collector, as shown in **Figure 6.9**:

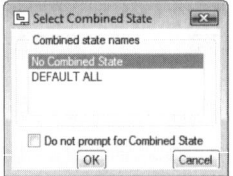

FIGURE 6.9

The **Select Combined State dialog box** provides two options:

- **No Combined State:** If you select the **No Combined State** option, then an unexploded view of the drawing view is placed in the Drawing file.
- **DEFAULT ALL:** If you select the **DEFAULT ALL** option, then an exploded view of the drawing view is placed in the Drawing file.

In our case, we want an unexploded view to be placed in the Drawing file. So, we select the **No Combined State** option.

> **Note:** The **Select Combined State dialog box** appears only when the selected model is an **Assembly**.

3. *Select* the **No Combined State** option from the **Combined state names collector** and *click* the **OK** button, as shown in **Figure 6.10**:

FIGURE 6.10

A prompt to select a **CENTER POINT** to place on the drawing view appears (**Figure 6.11**).

4. *Select* a center point on the **Drawing** area, as shown in Figure 6.11:

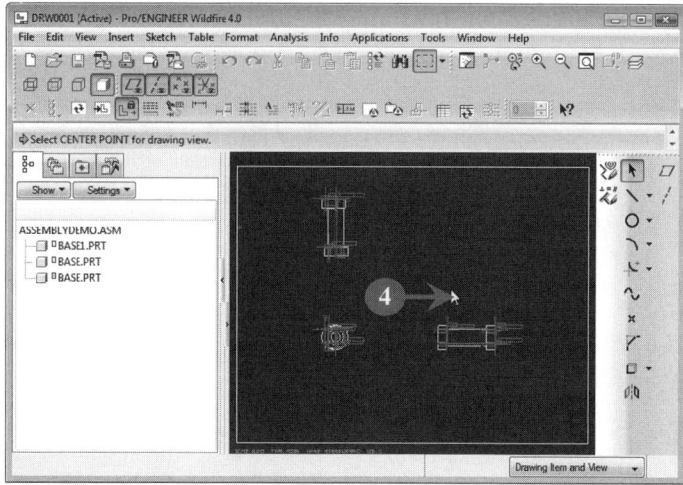

FIGURE 6.11

Once you select the **CENTER POINT**, the **Drawing View dialog box** also appears (**Figure 6.12**).

5. *Select* the **Standard Orientation** from the **Model view names** list box in the **Drawing View dialog box**, as shown in Figure 6.12:

6. *Click* the **Apply** button, as shown in Figure 6.12:

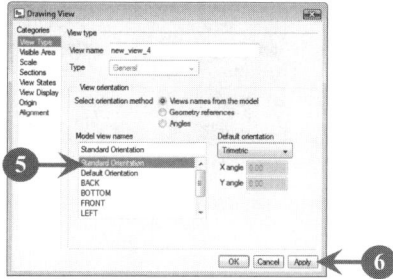

FIGURE 6.12

When you click the **Apply** button (Figure 6.12), the **Cancel** button changes to the **Close** button (**Figure 6.13**).

In the **Drawing View dialog box**, the **General** option is selected by default in the **Type** drop-down list. The **Views names from the model** radio button is selected by default in the **View orientation** area to display the various orientations to choose. Once the **Standard Orientation** option is selected from the **Model view names** list box, you need to *click* the **Close** button to apply the selected orientation on the **General** view.

7. Now, *click* the **Close** button to complete the process of generating the **General** view, as shown in Figure 6.13:

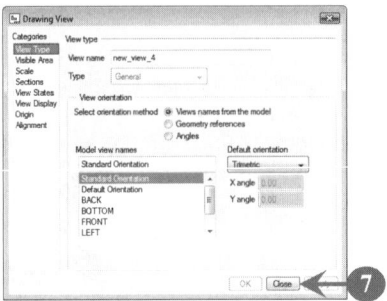

FIGURE 6.13

The **General** view of the selected model (ASSEMBLYDEMO.ASM) is generated, as shown in **Figure 6.14**:

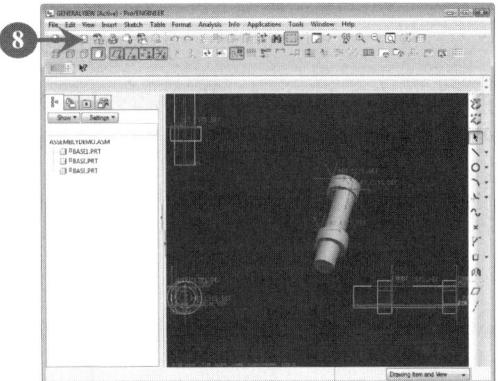

FIGURE 6.14

8. Now, save the file in the working directory by *clicking* the **Save** button (Figure 6.14) in the **File toolbar**.

Once the **General** view is generated, you can place the **Projection** view on it. The **General** view mostly acts as the parent view for the other views. Let's discuss the **Projection** view in the following section.

Projection View

A **Projection** view is an orthographic view generated from an existing drawing view. In an orthographic view, all objects appear at the same scale.

Before generating the **Projection** view, you must have at least one parent view present on the Drawing file. To generate a **Projection** view, you need to specify a location on the **Drawing Area** with respect to another drawing view. Pro/ENGINEER Wildfire 4.0 automatically determines how to project the generated view. Let's take a look at the steps required to generate a **Projection** view.

Perform the following steps to generate a **Projection** view:

1. *Start* a new Drawing file named PROJECTIONVIEW.DRW with the default model named ASSEMBLYDEMO.ASM. Since at least one parent view should be present before generating a **Projection** view, we will generate a **General** view first.

2. *Generate* the **General** view. A **Projection** view can now be placed in the Drawing file.

3. *Select* the **Insert > Drawing View > Projection** option from the **Menu Bar**, as shown in **Figure 6.15**:

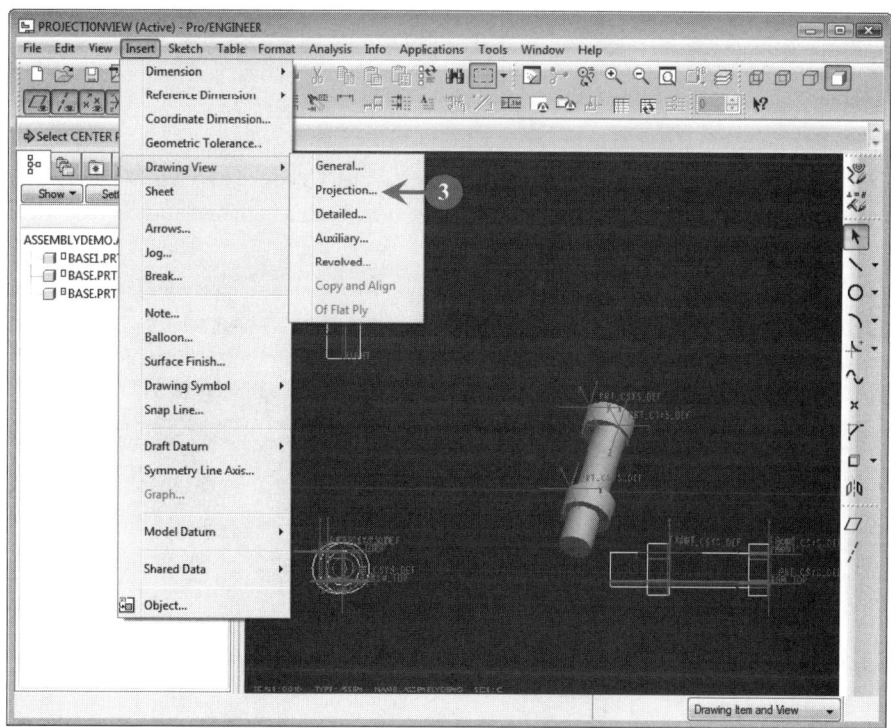

FIGURE 6.15

A box with a yellow color outline appears in the Drawing file (**Figure 6.16**).

4. *Move* the box to a location where you need to place the view, as shown in Figure 6.16:

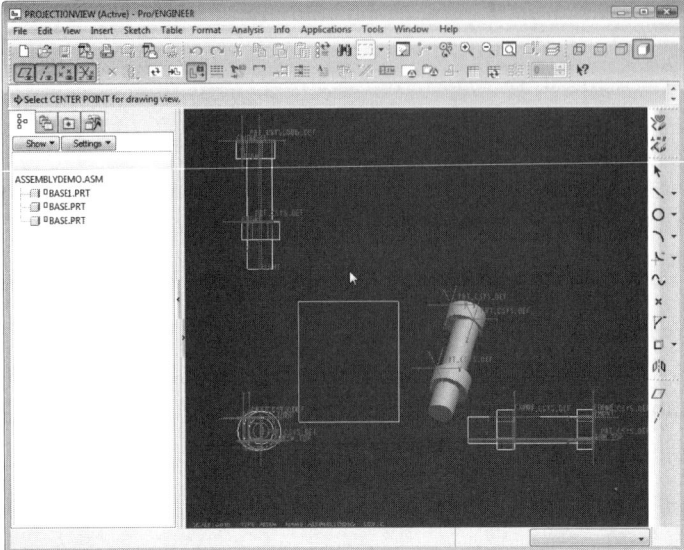

FIGURE 6.16

When you click on the desired location, the **Projection** view of the selected model (ASSEMBLYDEMO.ASM) is generated, as shown in **Figure 6.17**:

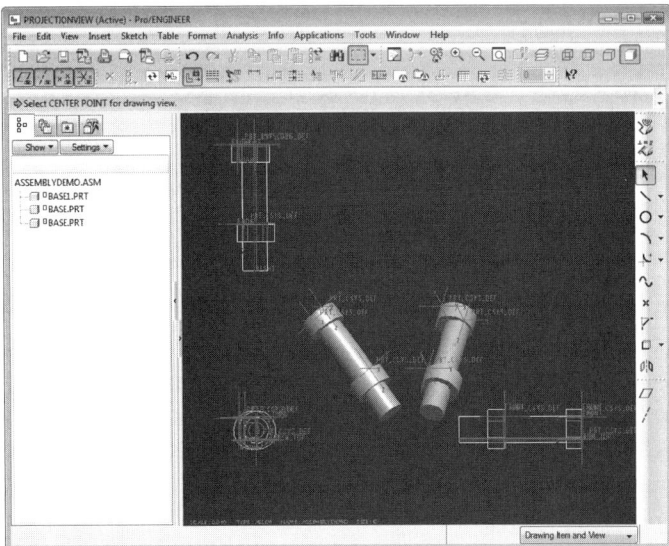

FIGURE 6.17

5. Now, save the file in the working directory. Once the **Projection** view has been generated, you can place other views on it.

Let's discuss the **Detailed** view in the following section.

Detailed View

A **Detailed** view is a view that displays a particular portion of an existing view in a larger viewing scale to help the user clearly and easily see the geometry and dimensions in the view. Dimensions are discussed in detail in the section "Inducing Dimensions and Tolerance" later in this chapter.

Let's look at the steps required to generate a **Detailed** view. Before generating the **Detailed** view, you must have at least one parent view present on the Drawing file. In our case, we will generate a **Detailed** view on the existing **General** view present in the Drawing file.

Perform the following steps to generate a **Detailed** view:

1. *Start* a new Drawing file named DETAILEDVIEW.DRW with the default model named as ASSEMBLYDEMO.ASM. Since at least one parent view should be present, we will first generate a **General** view before generating a **Detailed** view.

2. *Generate* the **General** view. *Select* the **Insert > Drawing View > Detailed** from the **Menu Bar**, as shown in **Figure 6.18**:

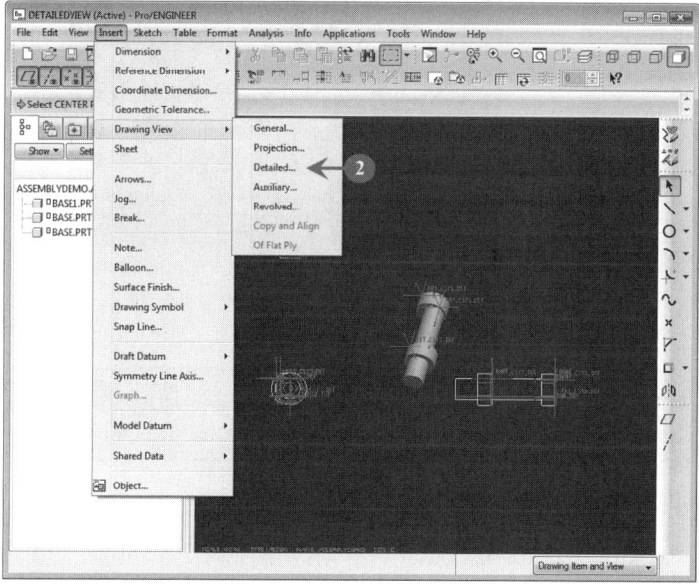

FIGURE 6.18

The **Select dialog box** appears to select a point for detail on an existing view, as shown in **Figure 6.19**:

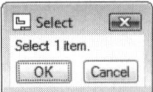

FIGURE 6.19

When you select a point on an existing view, the **Select dialog box** automatically gets closed.

You are prompted to sketch a spline around the point selected of an existing view that you want to enlarge.

3. *Draw* the spline in such a way that it covers the point selected, as shown in **Figure 6.20**:

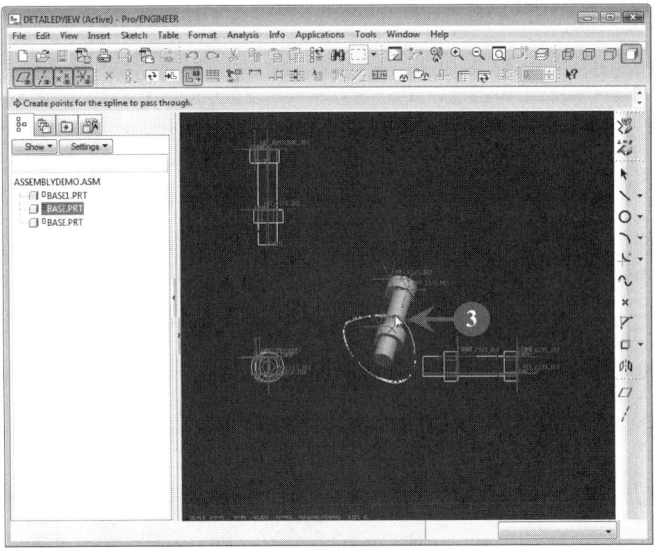

FIGURE 6.20

Note: Spline must not be drawn using the **Sketcher toolbar** located at the right side of the **Drawing Area**. *Click* the **Drawing Area** to start drawing the spline.

4. *Click* the middle **mouse** button to complete the spline and move the cursor to the location where you need to place the view, as shown in **Figure 6.21**:

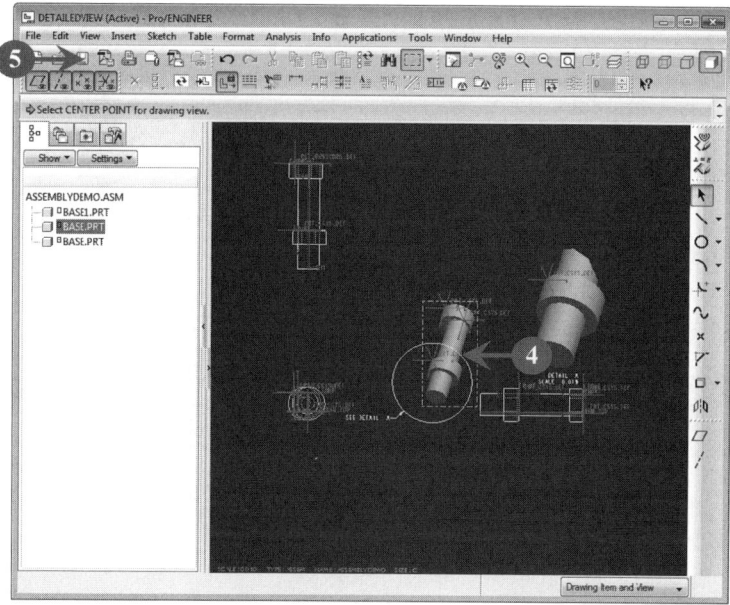

FIGURE 6.21

5. *Save* the file in the working directory.

Let's discuss the **Auxiliary** view in the following section.

Auxiliary View

An **Auxiliary** view is similar to the **Projection** view—the only difference is that the view is generated at right angles to a selected surface, datum plane, or a specified edge of an existing view. Before generating the **Auxiliary** view, at least one parent view must be present on the Drawing file. In our case, we consider the **Projection** view as the parent view. Let's generate an **Auxiliary** view on the existing **Projection** view present in the Drawing file.

Perform the following steps to generate an **Auxiliary** view:

1. *Start* a new Drawing file named AUXILIARYVIEW.DRW with the default model named as ASSEMBLYDEMO.ASM. Since at least one parent view should be present before generating an **Auxiliary** view, we first generate a **Projection** view.

2. *Generate* the **Projection** view. Now, an **Auxiliary** view can be placed on the Drawing file.

3.   *Select* the **Insert>Drawing View>Auxiliary** option from the **Menu Bar**, as shown in **Figure 6.22**:

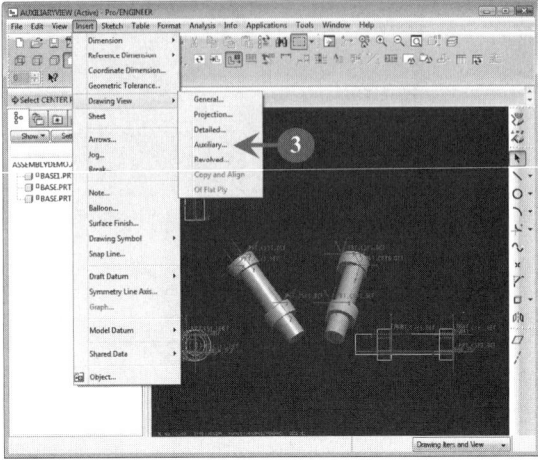

FIGURE 6.22

The **Select dialog box** appears to select an item (edge) on an existing view (Figure 6.19).

4.   *Select* the desired edge on the **Projection** view, as shown in **Figure 6.23**:

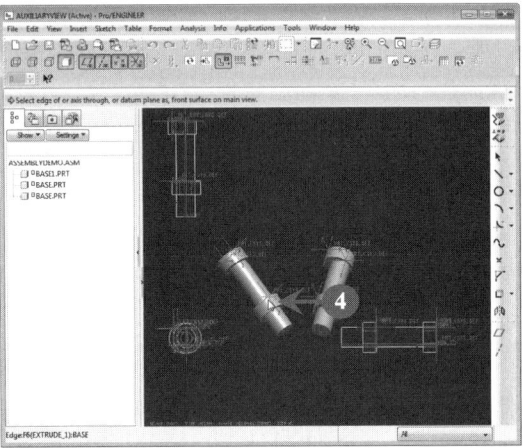

FIGURE 6.23

A box with a yellow color outline appears in the **Drawing Area**.

5.   *Move* the box to the location where you want to place the view, as shown in **Figure 6.24**:

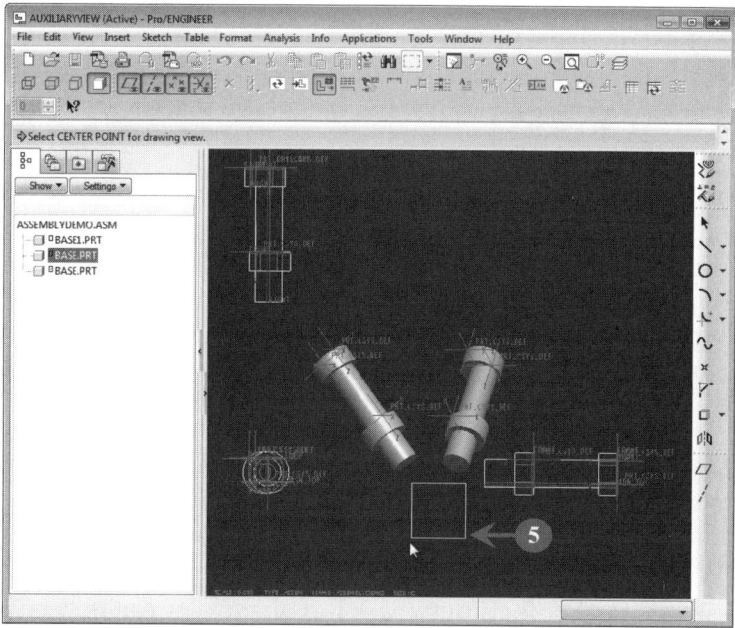

FIGURE 6.24

When you click on the desired location, the **Auxiliary** view of the selected view is generated, as shown in **Figure 6.25**:

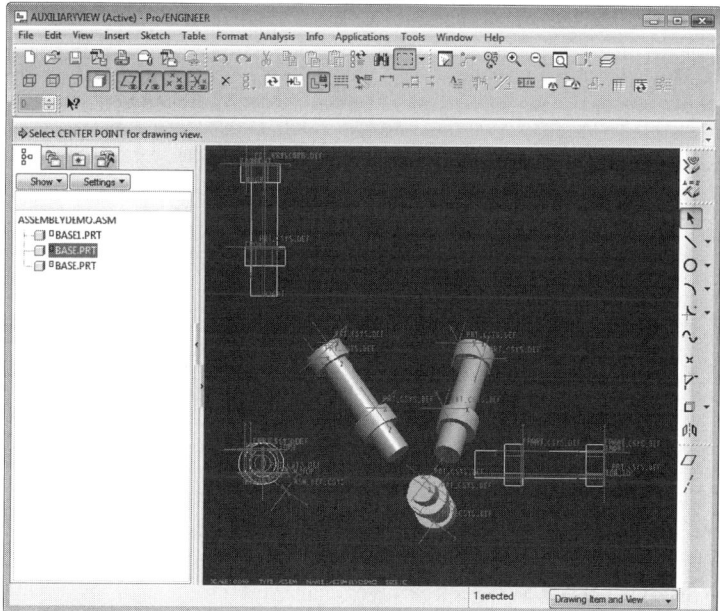

FIGURE 6.25

6. *Save* the file in the current directory.

Now, in the following section, let's discuss the **Revolved** view.

Revolved View

A **Revolved** view is that section of an existing view that has been revolved to 90° around a cutting plane. A cutting plane is a plane that shows the cutting pattern of a section. An area crosssection of the existing view cut by the cutting plane is known as a section.

At least one parent view must be present on the Drawing file before generating the **Revolved** view. In our case, we consider the **Projection** view as the parent view. Let's generate a **Revolved** view on the existing **Projection** view present in the Drawing file.

Perform the following steps to generate a **Revolved** view:

1. *Start* a new Drawing file named REVOLVEDVIEW.DRW with the default model named as ASSEMBLYDEMO.ASM. Since at least one parent view should be present before generating a **Revolved** view, we will first generate a **Projection** view.

2. *Generate* the **Projection** view. A **Revolved** view can now be placed on the Drawing file.

3. *Select* the **Insert > Drawing View > Revolved** option from the **Menu Bar**, as shown in **Figure 6.26**:

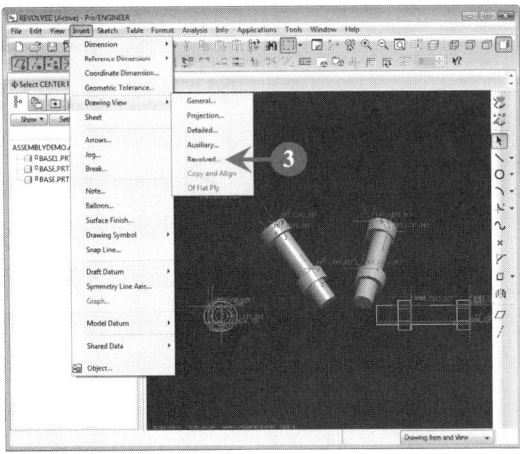

FIGURE 6.26

The **Select dialog box** appears to select a **Projection** view (Figure 6.19).

Once a **Projection** view on the **Drawing Area** is selected, a prompt to select a **CENTER POINT** is also displayed.

4. *Click* the center point in the **Drawing Area**. The **Drawing View dialog box** appears, as shown in **Figure 6.27**:

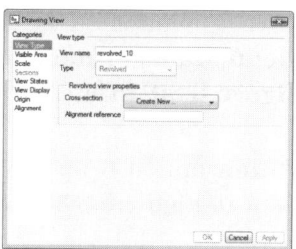

FIGURE 6.27

Apart from the **Drawing View dialog box**, the **Menu Manager menu** also appears (**Figure 6.28**).

5. *Select* **Planar > Single > Done** from the **XSEC CREATE** submenu, as shown in Figure 6.28:

FIGURE 6.28

A text box appears in the message area that prompts you to specify a name for the crosssection (**Figure 6.29**).

6. *Specify* the name of the crosssection in the **Message Input window** (Figure 6.29). We specified crosssection as the name, as shown in Figure 6.29:

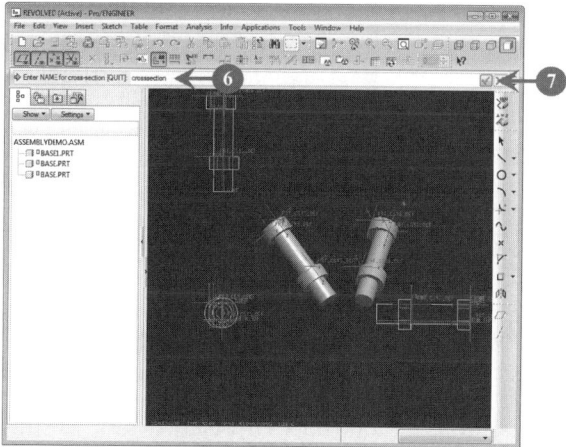

FIGURE 6.29

7. *Click* the **Build feature** (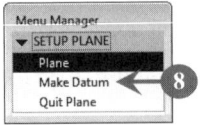) button on the **Message Input window** (Figure 6.29). The **Menu Manager menu** with the **SETUP PLANE** submenu appears (**Figure 6.30**) to either select or create a datum plane on the selected view.

8. *Select* the **Make Datum** option from the **SETUP PLANE** submenu to create a datum, as show in Figure 6.30:

FIGURE 6.30

The **DATUM PLANE** submenu appears under the **Make Datum menu** (**Figure 6.31**).

9. *Click* the **Normal** option from the **DATUM PLANE** to create a datum plane normal or perpendicular to the selected view, as shown in Figure 6.31:

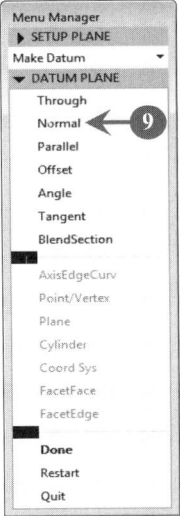

FIGURE 6.31

The **Select dialog box** appears to select a point (edge) on a selected view (**Figure 6.19**).

10. *Select* the desired edge on the **Projection** view, as shown in **Figure 6.32**:

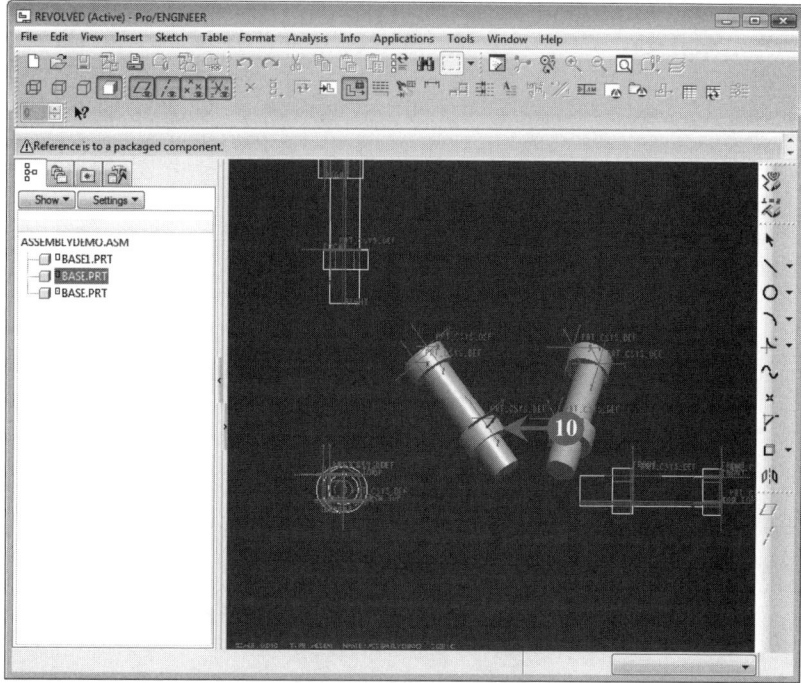

FIGURE 6.32

11. *Select* the **Done** option in the **Menu Manager menu** to finish the creation of the datum plane, as shown in **Figure 6.33**:

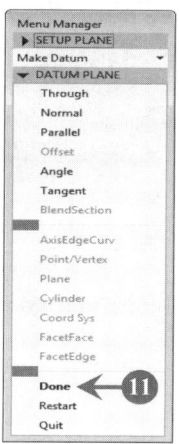

FIGURE 6.33

The **DATUM PLANE** submenu disappears.

12. *Click* the **Apply** button in the **Drawing View dialog box**, as shown in **Figure 6.34**:

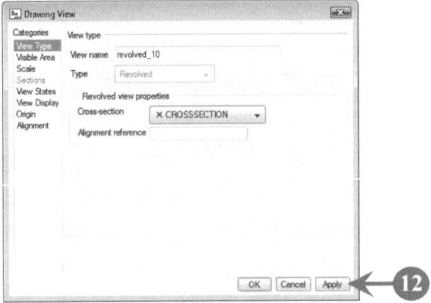

FIGURE 6.34

13. *Click* the **Close** button to exit the **Drawing View dialog box**. The **Revolved** view of the selected model (ASSEMBLYDEMO.ASM) is generated, as shown in Figure 6.35:

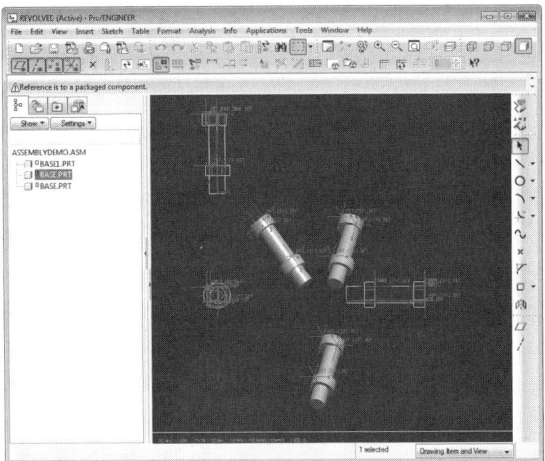

FIGURE 6.35

14. *Save* the file in the working directory.

Till now, we discussed the drawing views supported by Pro/ENGINEER Wildfire 4.0 such as **General** view, **Projection** view, **Detailed** view, **Auxiliary** view, and **Revolved** view. These drawing views are required to convey all the necessary information about a component so that the component can be manufactured easily. A **General** view must be the first view that should be placed in the Drawing file so that other drawing views can be generated on top of it.

Next, let's discuss modifying these drawing views.

6.3 MODIFYING THE DRAWING VIEWS

Pro/ENGINEER Wildfire 4.0 allows the user to easily modify existing drawing views. Most of the options available for modifying drawing views are selected from the context menu, which is displayed when you right-click a drawing view.

Pro/ENGINEER Wildfire 4.0 allows the user to perform the following modifications on the drawing views:

- Changing the type of a view
- Changing the scale of a view
- Reorienting the view
- Moving a view
- Erasing a view
- Deleting a view

Now, let's discuss each of these modifications.

Changing the Type of a View

There are several drawing views in Pro/ENGINEER Wildfire 4.0 such as **General** view, **Projection** view, **Detailed** view, **Auxiliary** view, and **Revolved** view. Pro/ENGINEER Wildfire 4.0 allows the user to change the type of any existing drawing view, except the **Detailed** view. Let's take a look at the steps required to change the type of an existing view.

Perform the following steps to change the type of an existing view. In our case, we have selected the **Projection** view to change its type.

1. *Open* the PROJECTIONVIEW.DRW file and *right-click* the **Projection** view to invoke the context menu, as shown in Figure 6.36:

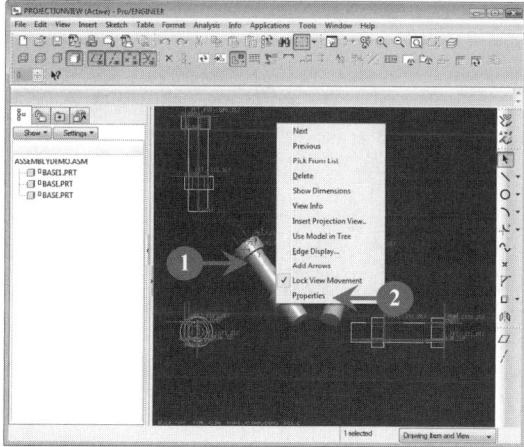

FIGURE 6.36

2. *Select* the **Properties** option from the context menu (Figure 6.36). The **Drawing View dialog box** appears and displays various options for modifying the view (Figure 6.37).

3. *Select* the desired view type from the **Type** drop-down list in the **View Type option in the Categories** area. In our case, we have selected the **General** view, as shown in Figure 6.37:

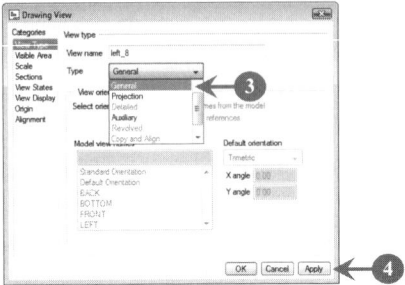

FIGURE 6.37

4. *Click* the **Apply** button to apply the selected view type to the current **Projection** view (Figure 6.37).

When you *click* the **Apply** button, the **Cancel** button changes to the **Close** button.

5. *Click* the **Close** button to exit the **Drawing View dialog box**. The selected **Projection** view is successfully changed into **General** view, as shown in **Figure 6.38**:

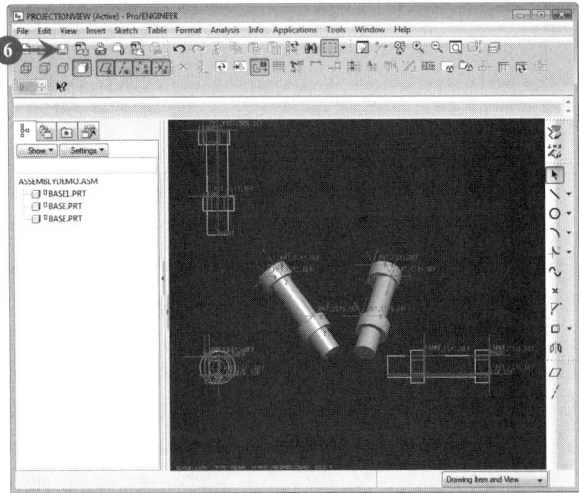

FIGURE 6.38

6. *Save* the file in the working directory.

In the next section, we discuss changing the viewing scale of a drawing view.

Note: Users can modify the view type of only the **General**, **Projection**, and **Auxiliary** view. These views are also used to change the type for the view, which has been selected to get modified.

Changing the Scale of a View

Pro/ENGINEER Wildfire 4.0 uses the drawing scale to control the scale of an existing view. In Pro/ENGINEER Wildfire 4.0, users are allowed to modify and change the scale of either an existing **General** view or a **Detailed** view only. Users can change the scale of an existing view using the **Drawing View dialog box**. Let's take a look at the steps required to change the scale of an existing view.

Perform the following steps to change the scale of an existing view:

1. *Right-click* the desired view in the PROJECTIONVIEW.DRW file. A context menu appears, as shown in **Figure 6.39**:

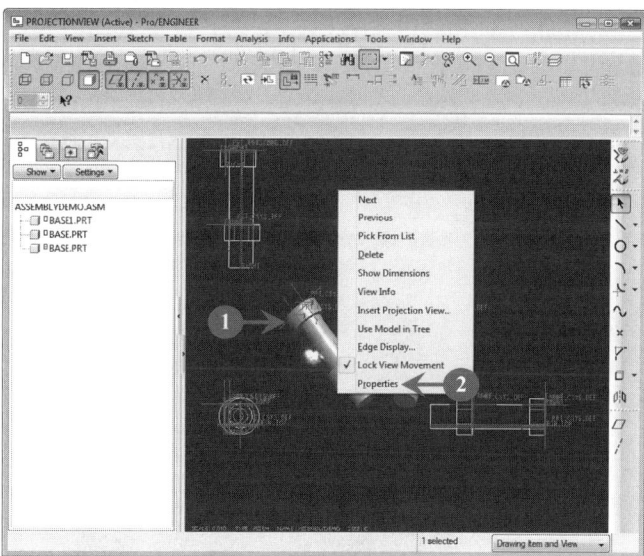

FIGURE 6.39

2. *Select* the **Properties** option from the shortcut menu (Figure 6.39). The **Drawing View dialog box** appears and displays various options for modifying the view (**Figure 6.40**).

3. *Select* the **Custom scale** radio button from the **Scale menu** under the **Categories** area to specify a value (Figure 6.40). In our case, 0.020 is specified as the scale value.

4. *Click* the **Apply** button to confirm the specified value, as shown in Figure 6.40:

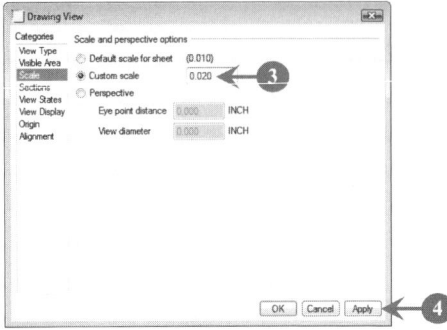

FIGURE 6.40

When you click the **Apply** button, the **Cancel** button changes to the **Close** button.

5. *Click* the **Close** button to exit the **Drawing View dialog box** (Figure 6.40). The view scale of the selected **General** view is successfully changed, as shown in **Figure 6.41**:

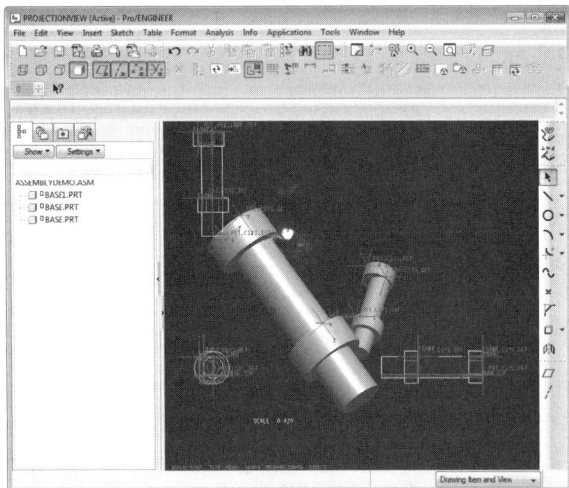

FIGURE 6.41

6. *Save* the file in the working directory.

Now, let's discuss changing the orientation of a drawing view in the following section.

Reorienting the View

Pro/ENGINEER Wildfire 4.0 allows users to change the orientation of an existing view using the **Drawing View dialog box**. Let's take a look at the steps required to change the orientation of an existing view.

In our case, we have selected the **General** view to change its orientation. Perform the following steps to change the orientation of an existing view:

1. *Right-click* the **General** view in the PROJECTIONVIEW.DRW file and open the **Drawing View dialog box**, as shown in **Figure 6.42**:

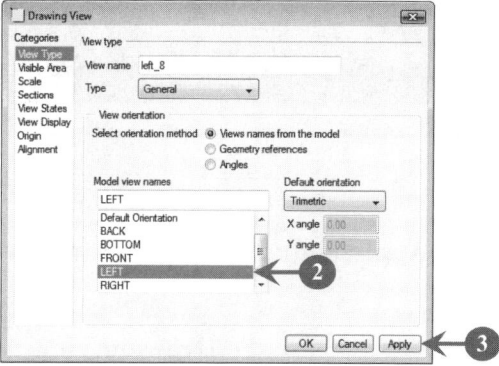

FIGURE 6.42

The **Drawing View dialog box** appears and displays various options for changing the orientation of the view. The **standard** orientation is the option that is selected by default under the **Model view names menu** in the **Drawing View dialog box**.

2. *Select* the **LEFT** option from the **Model view names** collector to change the orientation of the selected view (**Figure 6.42**).

3. *Click* the **Apply** button (Figure 6.42). The orientation of the selected view is successfully changed from **Standard** to **LEFT** orientation, as shown in **Figure 6.43**:

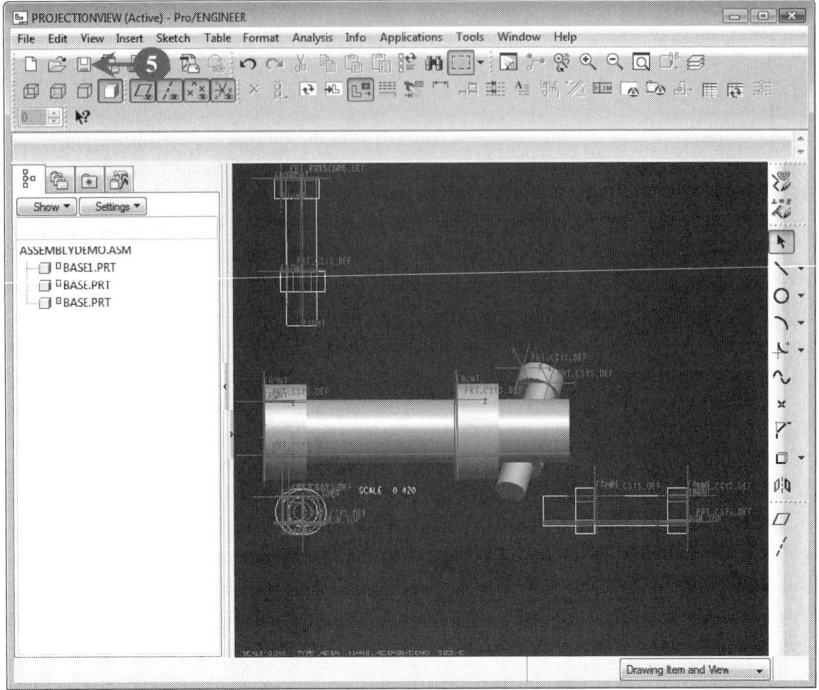

FIGURE 6.43

When you click the **Apply** button, the **Cancel** button changes to the **Close** button.

4. *Click* the **Close** button to exit the **Drawing View dialog box**.
5. *Save* the file in the working directory.

In the following section, let's discuss moving a drawing view from one location to another.

Moving a View

Pro/ENGINEER Wildfire 4.0 allows the user to easily move a view from one location to another in a Drawing file. However, if a parent view is moved, then the child views also move accordingly. Let's take a look at the steps required to move an existing view.

In our case, we have selected the **General** view to move a view. Perform the following steps to move an existing view:

1. *Right-click* the desired view in the PROJECTIONVIEW.DRW file. A context menu appears, as shown in **Figure 6.44**:

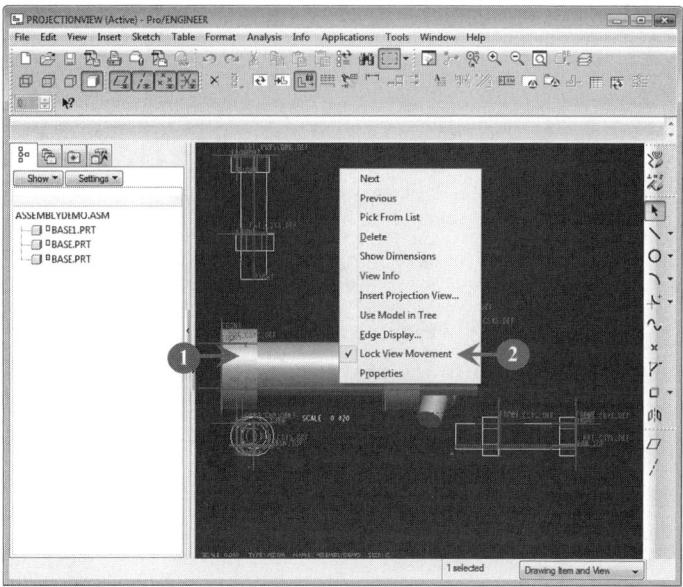

FIGURE 6.44

2. *Uncheck* the **Lock View Movement** option from the context menu (Figure 6.44). The selected view gets unlocked. Now the view can be moved to a desired location in the Drawing file.

3. *Drag* the selected view to a desired location, as shown in **Figure 6.45**:

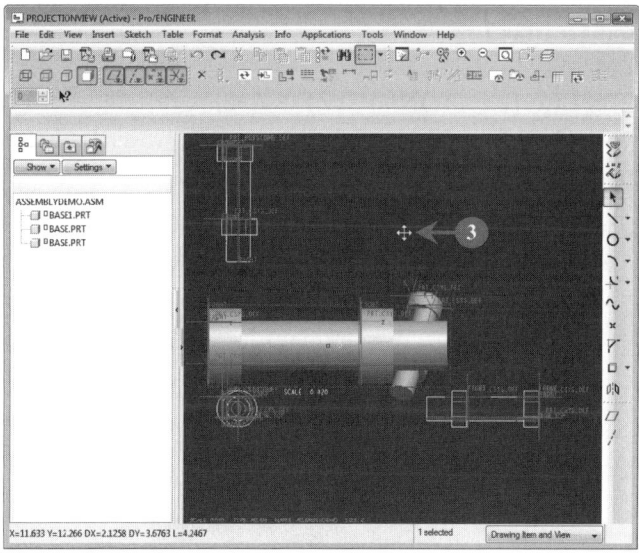

FIGURE 6.45

The selected **General** view is successfully moved from its original location, as shown in **Figure 6.46**:

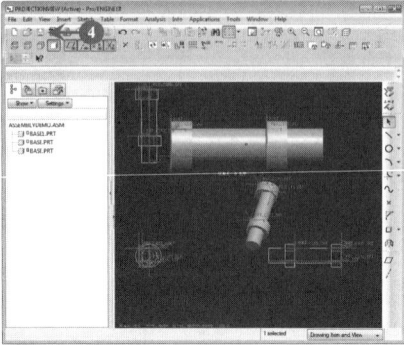

FIGURE 6.46

4. *Save* the file in the working directory.

Next, let's discuss erasing a drawing view in the following section.

Erasing a View

Pro/ENGINEER Wildfire 4.0 allows the user to erase a view from the Drawing file. Erasing a view temporarily removes the selected view from the Drawing file, but the erased view remains in the memory of the Drawing file. Let's take a look at the steps required to erase the drawing view.

In our case, we have selected the **General** view to erase. Perform the following steps to erase the selected **General** view:

1. *Open* the PROJECTIONVIEW.DRW file and Select the **View > Drawing Display > Drawing View Visibility** option from the **Menu Bar**, as shown in **Figure 6.47**:

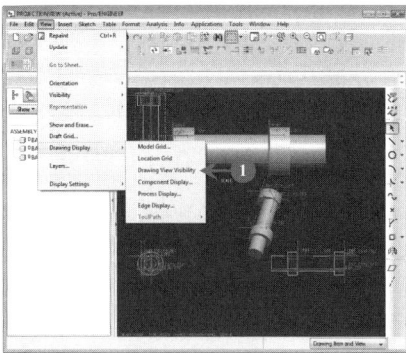

FIGURE 6.47

The **Menu Manager menu** appears, as shown in **Figure 6.48**:

FIGURE 6.48

Apart from the **Menu Manager menu**, the **Select dialog box** also appears to select an existing view to erase, as shown in **Figure 6.49**:

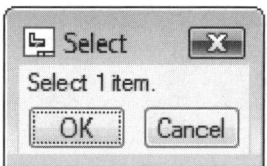

FIGURE 6.49

2. *Select* the desired view in the Drawing file to erase, as shown in **Figure 6.50**:

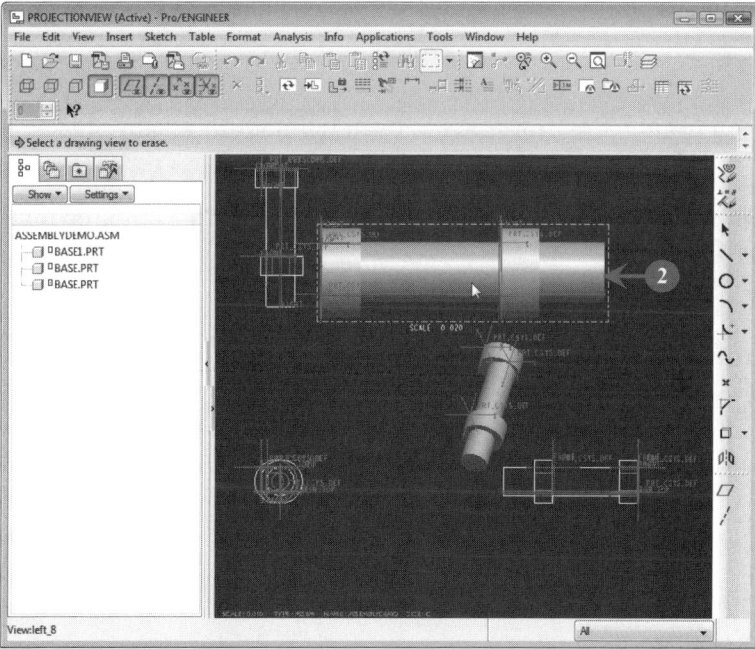

FIGURE 6.50

The selected view is erased and a box appears in place of the erased view (**Figure 6.51**).

3. *Click* the **OK** button in the **Select dialog box** to confirm the action performed (**Figure 6.49**).

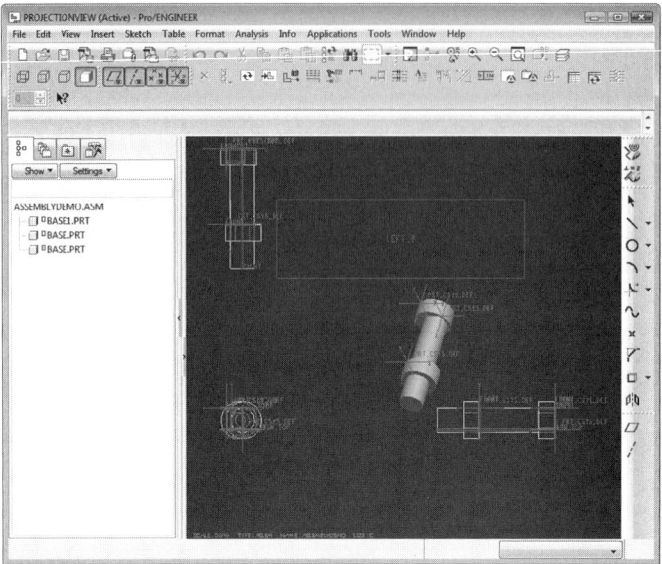

FIGURE 6.51

You can resume the view that has been temporarily removed from the Drawing file by performing the following steps:

1. *Select* the **Resume View** option from the **Menu Manager menu**, as shown in **Figure 6.52**:

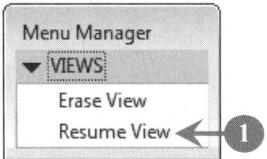

FIGURE 6.52

The **Select dialog box** appears (Figure 6.49) to select the box representing the erased view.

2. *Select* the box representing the erased view in the Drawing file to resume, as shown in **Figure 6.53**:

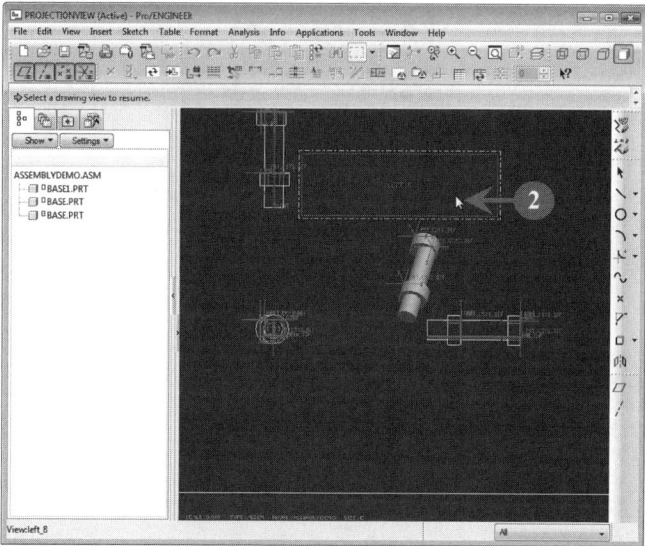

FIGURE 6.53

3. *Select* the **Done Sel** option from the **Menu Manager menu**, as shown in **Figure 6.54**:

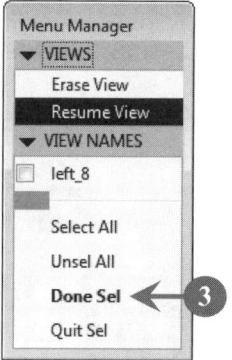

FIGURE 6.54

The erased view is successfully resumed in the Drawing file, as shown in **Figure 6.55**:

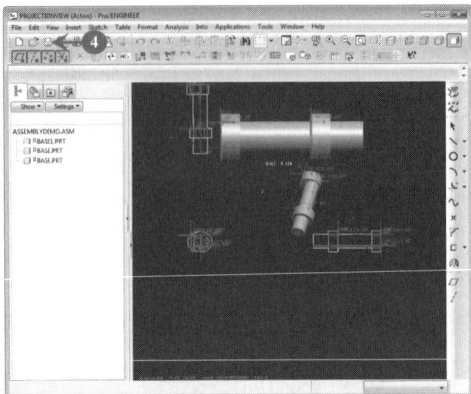

FIGURE 6.55

Note: The **Menu Manager menu** remains unclosed.

---

4. *Save* the file in the working directory.

Let's discuss deleting a drawing view in the following section.

Deleting a View

Pro/ENGINEER Wildfire 4.0 allows the user to delete a view from the Drawing file. Deleting a view permanently removes the selected view from the memory of the drawing. Let's take a look at the steps required to delete the drawing view.

Perform the following steps to delete the selected **General** view:

1. *Open* the PROJECTIONVIEW.DRW file and *right-click* the desired view. A context menu appears, as shown in **Figure 6.56**:

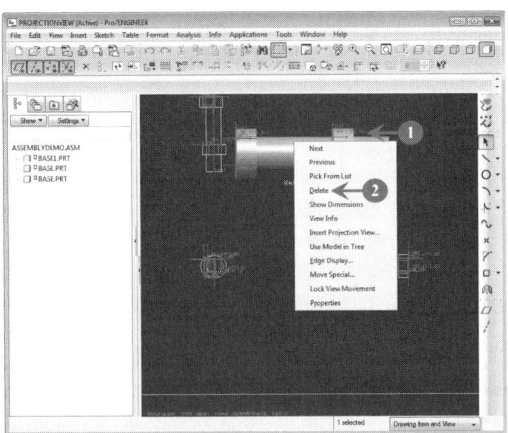

FIGURE 6.56

2. *Select* the **Delete** option from the context menu (Figure 6.56). The selected view is successfully deleted from the Drawing file, as shown in **Figure 6.57**:

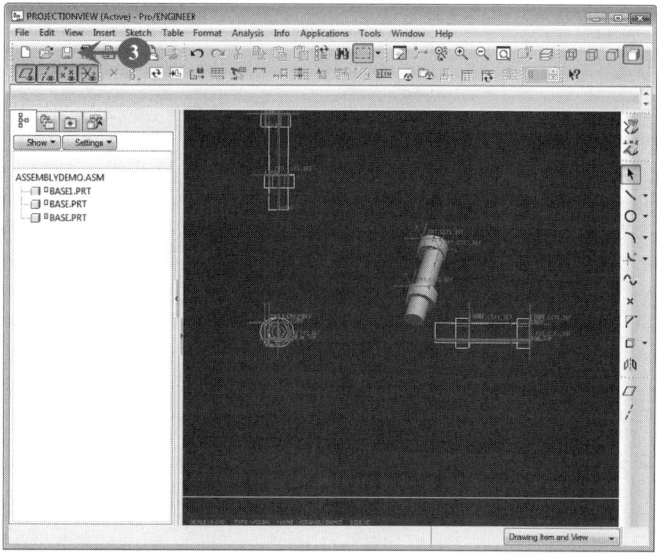

FIGURE 6.57

3. *Save* the file in the working directory.

Till now, we have discussed about modifying the existing drawing views. Let's now discuss adding dimensions and tolerances to the drawing views.

6.4 INDUCING DIMENSIONS AND TOLERANCE

Dimensions are the minimum number of coordinates that are required to specify every point in the drawing views. Pro/ENGINEER Wildfire 4.0 allows users to generate dimensions while the drawing views are being generated. Dimensions are assigned to the features in the selected view. These dimensions are associative in nature. This associative nature ensures that any modifications made in the dimensions in the Drawing file automatically updates its Part file. Apart from adding dimensions, tolerances are also induced in the dimensions. It is difficult to manufacture a component with exact dimensions. Tolerances are values that are added to the dimensions to handle the variation that occurs at the time of manufacturing.

Let's discuss inducing dimensions in the Drawing file.

In our case, we have selected the **Projection** view to induce dimensions. Perform the following steps to induce dimensions in the Drawing file:

1. *Create* a **Projection** view in the Drawing file, as shown in **Figure 6.58**:

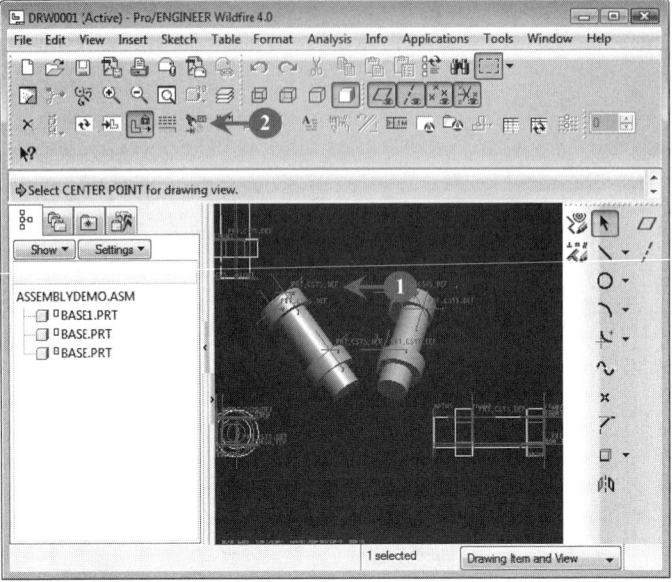

FIGURE 6.58

Note: The process to create a **Projection** view has been already discussed in the section, "Generating Drawing Views" of this chapter.

2. *Click* the **Show and Erase** (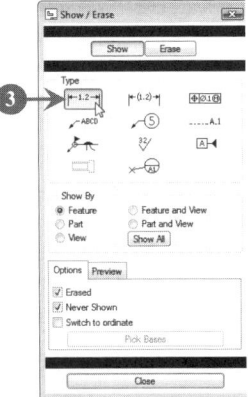) button from the **Menu Bar** (Figure 6.58). The **Show and Erase dialog box** appears and displays various options for dimensioning the view (**Figure 6.59**).
3. *Click* the **Dimension** button in the **Type** area in the **Show and Erase dialog box**, as shown in Figure 6.59:

FIGURE 6.59

The **Select dialog box** appears to select a feature in a **Projection** view, as shown in **Figure 6.60**:

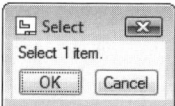

FIGURE 6.60

4. *Select* a feature in the **Projection** view to add a dimension, as shown in **Figure 6.61**:

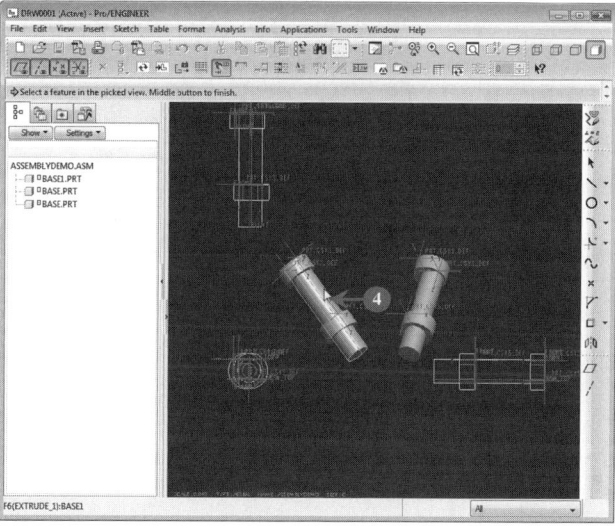

FIGURE 6.61

5. *Press* the middle **mouse** button and *click* the **Accept All** button in the **Show and Erase dialog box**, as shown in **Figure 6.62**:

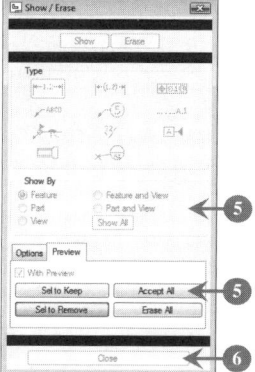

FIGURE 6.62

When you click the **Accept All** button, the **Close** button becomes enabled.

6.   Now, *Click* the **Close** button in the **Show and Erase dialog box** (Figure 6.62).

All the dimensions related to the selected feature are displayed in all views, as shown in **Figure 6.63**:

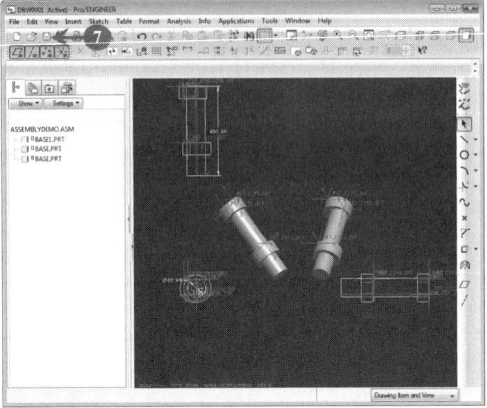

FIGURE 6.63

7.   *Save* the file in the working directory.

Let's discuss the inducing of tolerances in the dimensions of an existing view.

Perform the following steps to induce tolerance in the dimensions of an existing view:

1.   *Open* the DRW0001.DRW file and *select* the **File > Properties** option from the **Menu Bar**, as shown in **Figure 6.64**:

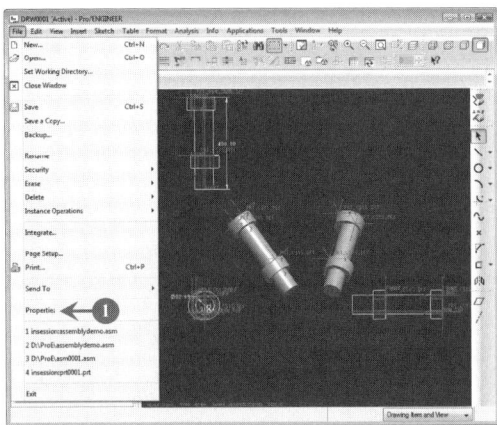

FIGURE 6.64

The **Menu Manager menu** appears and displays various options for manipulating the Pro/ENGINEER Wildfire 4.0 Drawing file properties.

2. *Select* the **Drawing Options** option from the **Menu Manager menu**, as shown in **Figure 6.65**:

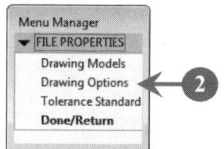

FIGURE 6.65

The **Options dialog box** appears and displays various drawing options to manipulate (**Figure 6.66**).

3. Specify **tol_display** in the **Option** edit box and specify "yes" in the **Value** edit box, as shown in **Figure 6.66**:

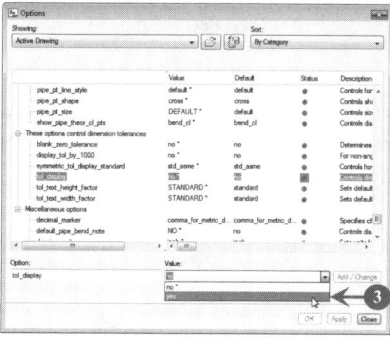

FIGURE 6.66

4. *Click* the **Add/Change** button in the **Options dialog box** to add the specified value of the **tol_display** option, as shown in **Figure 6.67**:

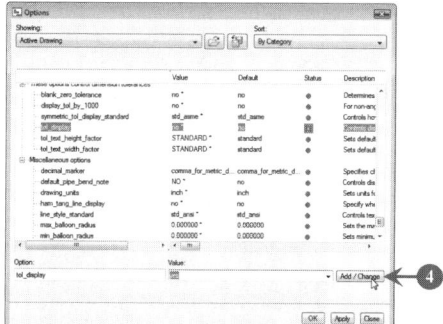

FIGURE 6.67

When the **tol_display** option is set to "yes", and the specified value is added, the display of tolerances for the current session is enabled.

5. *Click* the **Apply** button in the **Options dialog box** to set the specified value **tol_display** option, as shown in **Figure 6.68**:

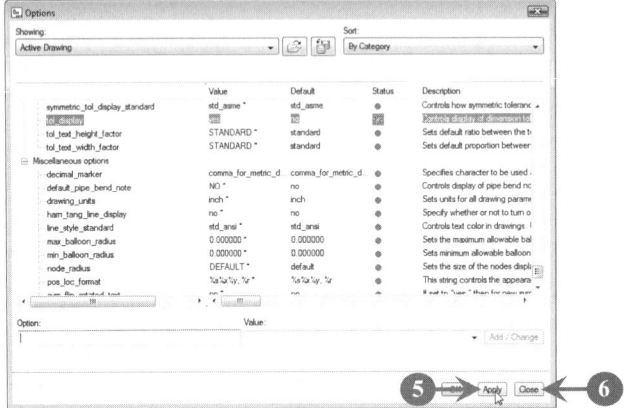

FIGURE 6.68

6. *Click* the **Close** button to exit the **Options dialog box** (Figure 6.68).
7. *Select* the **Done/Return** option from the **FILE PROPERTIES** submenu in the **Menu Manager menu**, as shown in **Figure 6.69**:

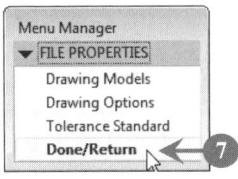

FIGURE 6.69

Geometric tolerances are induced successfully in the dimensions of the selected view, as shown in **Figure 6.70**:

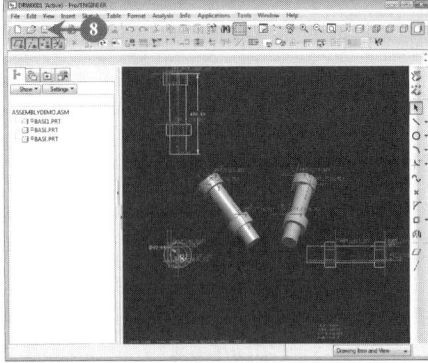

FIGURE 6.70

8. Finally, *save* the file in the working directory.

Now that the dimensions have been added to the drawing views, let's next discuss modifying them.

Modifying Dimensions

After adding dimensions in the drawing view, you may want to modify these dimensions by either changing the position of the text associated with the dimension (to make it readable) or changing the values of the dimensions. Pro/ENGINEER Wildfire 4.0 allows users to modify existing dimensions, which are added to specify every point in the drawing view. Let's perform the following steps to modify the dimensions of an existing view:

1. *Open* the DRW0001.DRW file and *select* the desired dimension (**Figure 6.71**).

2. *Select* the **Edit > Properties** from the **Menu Bar**, as shown in Figure 6.71:

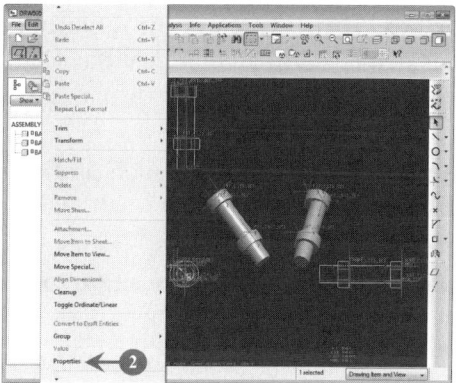

FIGURE 6.71

The **Dimension Properties dialog box** appears and the **Properties** tab is selected by default, as shown in **Figure 6.72**:

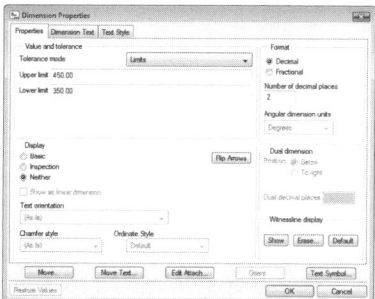

FIGURE 6.72

The **Dimension Properties dialog box** provides various options for modifying dimensions. Let's discuss the options present in the **Dimension Properties dialog box**:

- **Properties Tab:** The **Properties** tab contains various options for manipulating the properties of the dimensions. The **Properties** tab contains the following options:
  - **Value and Tolerance Area:** Provides an option named **Tolerance** mode for manipulating tolerance in the drawing view. The **Tolerance** mode drop-down list allows the user to select the type of **Tolerance** mode such as **Limits**, **Nominal**, and so on. The other options in the **Value and tolerance** area changes with the options selected in the **Tolerance** mode drop-down.
  - **Display Area:** Provides radio buttons such as **Basic**, **Inspection**, and **Neither** for providing the **Display Area** for the dimensions to be placed. The **Basic** radio button allows the user to display the dimension in a rectangular box. If the **Inspection** radio button is selected, then the dimension is displayed in an elliptical box. The **Neither** radio button is selected when you do not want to display the dimension in neither the **Basic** nor the **Inspection** type.
  - **Format Area:** Provides options such as **Decimal** and **Fractional** to display dimensions in decimal and fractional format.
  - **Dual Dimension Area:** Provides options such as **Position** and **Dual decimal places** for specifying a position to place dual dimensions and to place the number of decimal places in dual dimensions.
- **Dimension Text Tab:** The **Dimension Text** tab contains various options related to text associated with the dimensions, as shown in **Figure 6.73**:

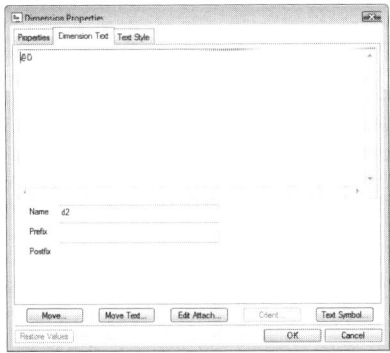

FIGURE 6.73

The **Dimension Text** tab contains:

- **Set Dimension Text Box:** Displays the text of the selected dimension.
- **Name Edit Box:** Displays the symbol associated with the dimension.
- **Prefix Edit Box:** Allows users to add additional information as a prefix to the default text associated with the dimension.
- **Postfix Edit Box:** Allows users to add additional information as a postfix to the default text associated with the dimension.
- **Text Style Tab:** The **Text Style** tab contains various styling options for the text associated with the dimensions, as shown in **Figure 6.74**:

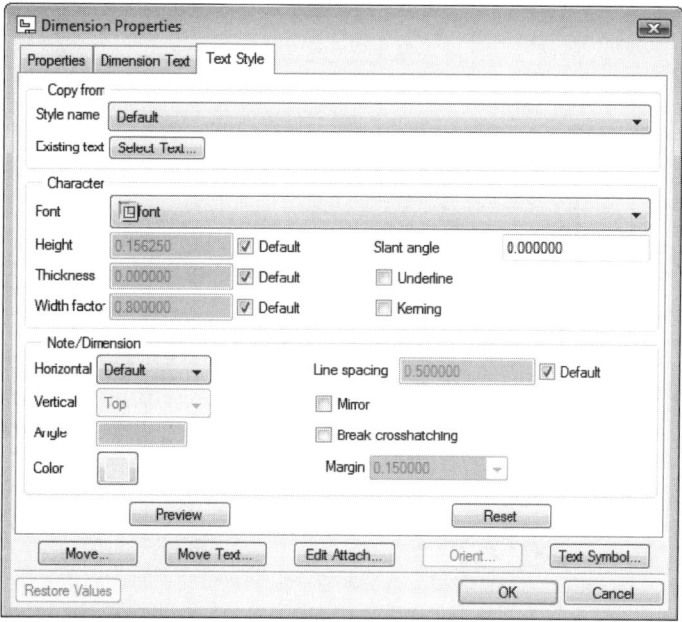

FIGURE 6.74

The **Text Style** tab contains the following options:

- **Copy From Area:** Provides options such as **Style name** and **Existing text** to modify the style of the text associated with the dimension.
- **Character Area:** Provides options such as **Font**, **Height**, **Thickness**, and so on to modify the font style of the text associated with the dimension.
- **Note/Dimension Area:** Provides options to modify the parameters associated with the dimensions such as color, line spacing, angle, and so on in the form of a note.
- **Preview Button:** Allows users to be able to see the preview of the changes made using the **Dimension Properties dialog box**.
- **Reset Button:** Allows users to reset the default values.

In our case, we will select the **Properties** tab for modifying dimensions.

3. *Specify* the new values in the **Upper** and **Lower limit** edit boxes, as shown in Figure 6.75. In our case, the upper limit is 500.00 and the lower limit is 400.00.

4. *Click* the **OK** button, as shown in Figure 6.75:

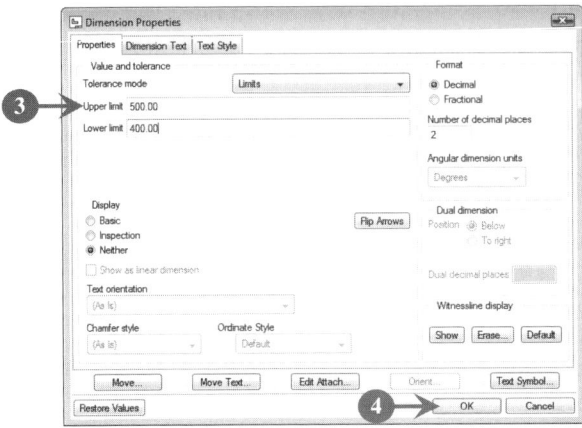

FIGURE 6.75

The **Upper** and **Lower limits** are values representing the highest and lowest size of the selected dimension. The limits associated with the selected dimension are modified, as shown in **Figure 6.76**:

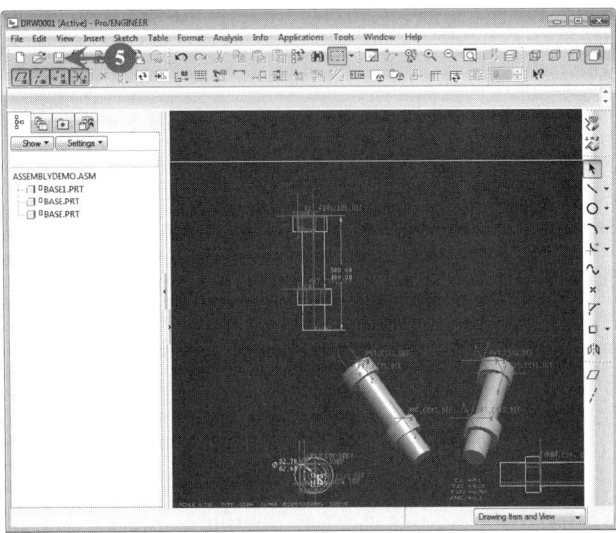

FIGURE 6.76

5. Now, **save** the file in the working directory.

Till now, we discussed the addition of dimensions and tolerances in drawing views. Apart from this, we also discussed the modification of existing dimensions. Next, let's discuss creating notes in the drawing views.

6.5 CREATING NOTES

Pro/ENGINEER Wildfire 4.0 allows users to provide additional information in the Drawing file by adding notes. The additional information can be specific to any **Part** or an **Entity** type. Occasionally, general information that does not apply to any particular part or feature is also specified in the note. The drawing notes contain information in textual as well as symbolic format. Let's discuss creating notes in a Drawing file.

Perform the following steps to create a note in a Drawing file:

1. *Open* the DRW0001.DRW file and *select* the **Insert > Note** option from the **Menu Bar**, as shown in **Figure 6.77**:

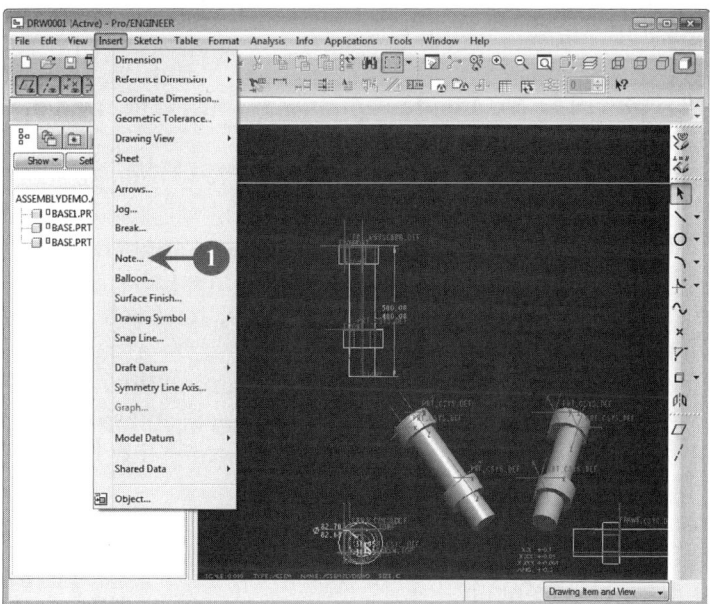

FIGURE 6.77

The **Menu Manager menu** appears and contains various options that facilitate the creation of notes in a Drawing file (**Figure 6.78**).

2. *Select* the **With Leader** option from the **Menu Manager menu**, as shown in Figure 6.78:

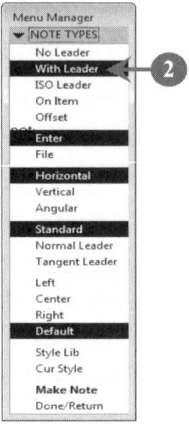

FIGURE 6.78

The **NOTE TYPES** submenu in the **Menu Manager** menu provides various options to specify the appearance of the drawing note. Let's discuss these options before we start working with them:

- **No Leader:** The **No Leader** option allows users to avoid the leader for the note to be placed in the Drawing file.
- **With Leader:** The **With Leader** option allows users to specify a leader for the note to be placed in a Drawing file.
- **ISO Leader:** The **ISO Leader** option allows users to specify an ISO-type leader. An ISO-type leader contains underlined text.
- **On Item:** The **On Item** option allows users to attach a note directly on the selected item in the Drawing file.
- **Offset:** The **Offset** option allows users to group the note with any selected entity such as dimension, geometric tolerance, and so on.
- **Enter:** The **Enter** option allows users to enter text for the note from the keyboard.
- **File:** The **File** option allows users to provide text for the note from a file.
- **Horizontal:** The **Horizontal** option allows users to create a note with the text placed horizontally.
- **Vertical:** The **Vertical** option allows users to create a note with the text placed vertically.
- **Angular:** The **Angular** option allows users to create a note with the text placed at a certain angle.
- **Standard:** The **Standard** option allows users to use the default leader type.

- **Normal Leader:** The **Normal Leader** option allows users to make the leader either normal or perpendicular to any entity type.
- **Tangent Leader:** The **Tangent Leader** option allows users to make the leader tangent to any entity type.
- **Left:** The **Left** option allows users to left align the text associated with the note.
- **Center:** The **Center** option allows users to centrally align the text associated with the note.
- **Right:** The **Right** option allows users to right align the text associated with the note.
- **Default:** The **Default** option allows users to align the note text with the Left alignment, which is the default alignment.
- **Style Lib:** The **Style Lib** option allows users to create a new style for the current session in which the user is currently working.
- **Cur Style:** The **Cur Style** allows users to select a new style for the current session from the list of predefined styles available in Pro/ENGINEER Wildfire 4.0.
- **Make Note:** The **Make Note** option allows users to start the process of creating a note in the Drawing file.

3. *Select* the **Make Note** option from the **Menu Manager** menu, as shown in **Figure 6.79**:

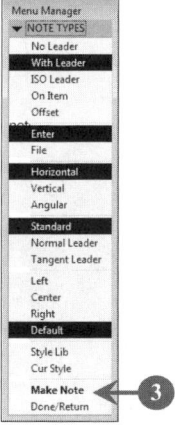

FIGURE 6.79

The **ATTACH TYPE** submenu appears in the **Menu Manager menu** (**Figure 6.80**).

4. *Select* the **On Surface** option from the **ATTACH TYPE** submenu to select a point on the surface of the front view to place a note, as shown in Figure 6.80:

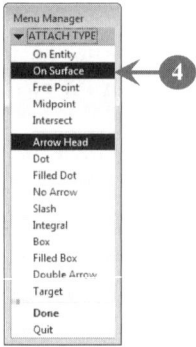

FIGURE 6.80

Apart from the **Menu Manager** menu, the **Select dialog box** also appears, as shown in **Figure 6.81**:

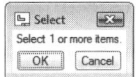

FIGURE 6.81

5. *Select* the desired surface from the front view in the Drawing file to place a note, as shown in **Figure 6.82**:

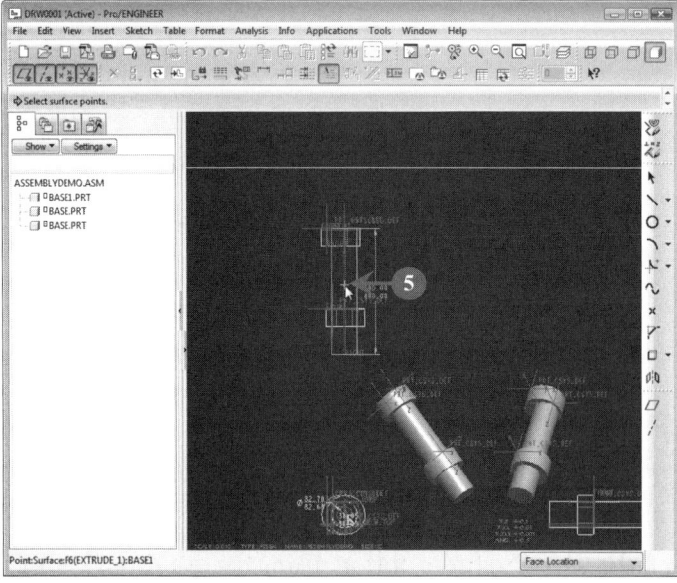

FIGURE 6.82

6. *Click* the selected surface in the **Drawing Area** to place a note. A red-colored box appears in place of the selected surface, as shown in **Figure 6.83**:

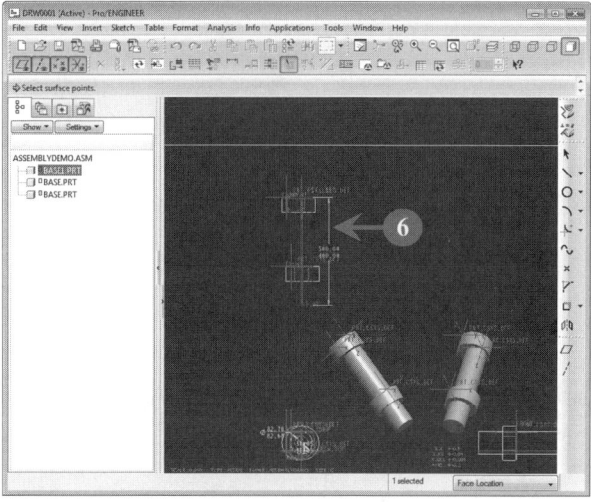

FIGURE 6.83

7. *Press* the middle **mouse** button in the middle of the red-colored box. The **Message Input window** appears to accept a note from the user, as shown in **Figure 6.84**:

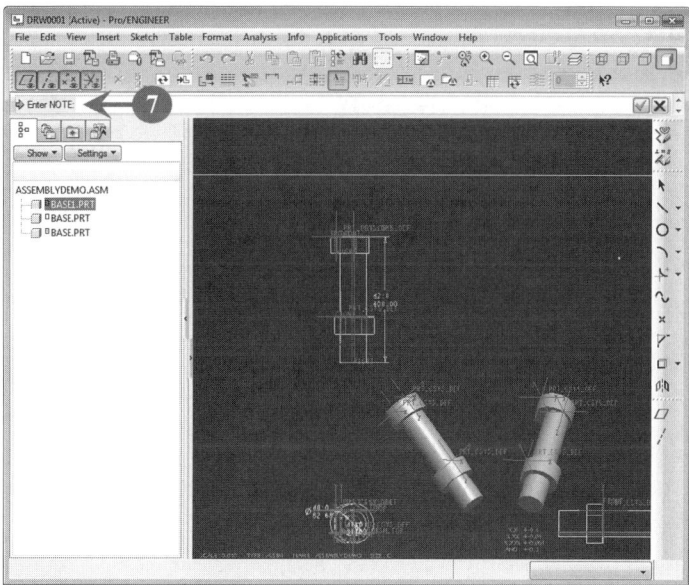

FIGURE 6.84

Apart from the **Message Input window**, the **Text Symbol dialog box** also appears, as shown in **Figure 6.85**:

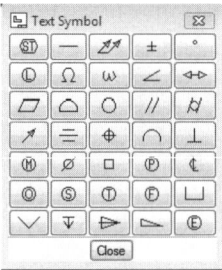

FIGURE 6.85

8. *Specify* a note in the **Message Input window**, as shown in **Figure 6.86**:

FIGURE 6.86

You can insert desired symbols in the note from the **Text Symbol dialog box**. In our case, we have selected the ▭ button to insert in the note.

9. *Click* the ▭ button in the **Text Symbol dialog box**, as shown in **Figure 6.87**:

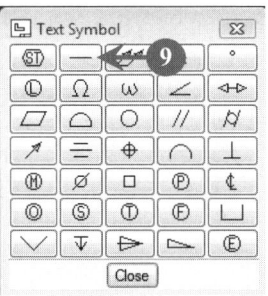

FIGURE 6.87

The selected symbol is appended to the text specified in the **Message Input window**, as shown in **Figure 6.88**:

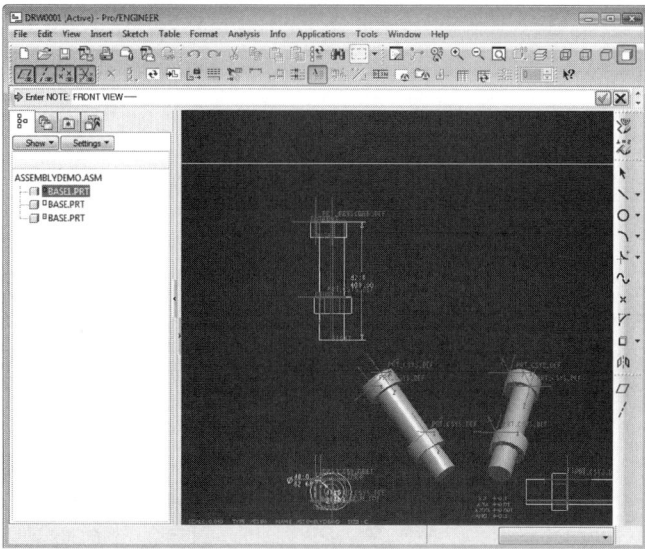

FIGURE 6.88

10. *Close* the **Text Symbol dialog box** and specify additional information in the **Message Input window**. Now, *click* the **Build feature** (☑) button, as shown in **Figure 6.89**:

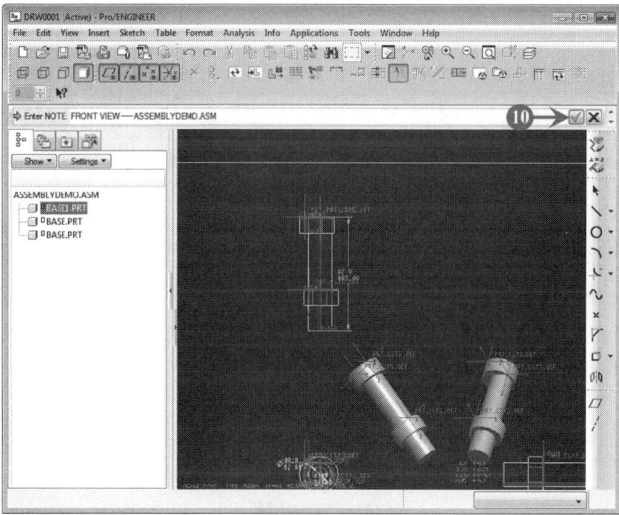

FIGURE 6.89

The **Drawing Area** previews the note that is created in the current session, as shown in **Figure 6.90**:

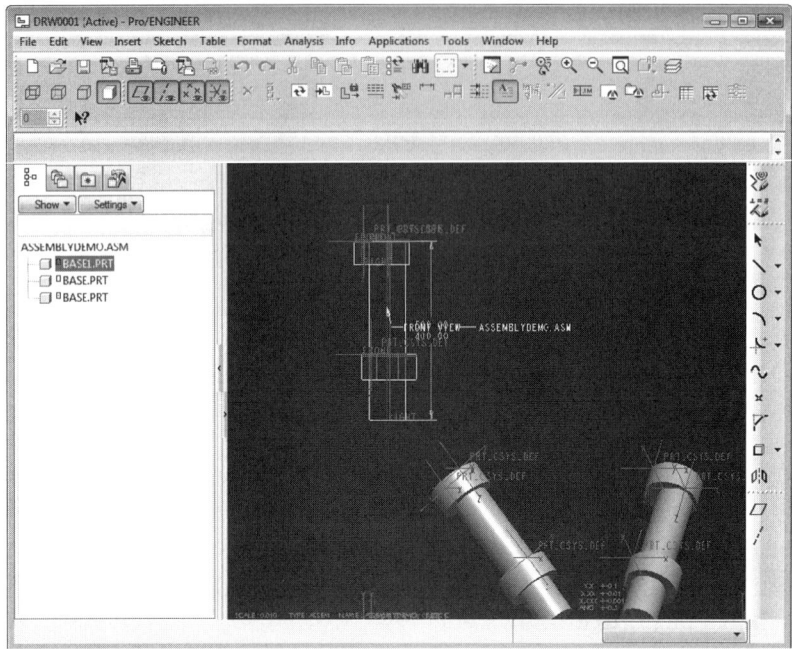

FIGURE 6.90

11. *Select* the **Done/Return** option from the **Menu Manager menu** to finish the note creation process and exit from the **Menu Manager menu**, as shown in **Figure 6.91**:

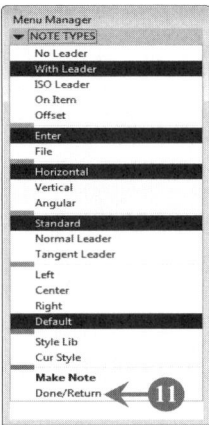

FIGURE 6.91

The note is placed successfully in the selected view, as shown in **Figure 6.92**:

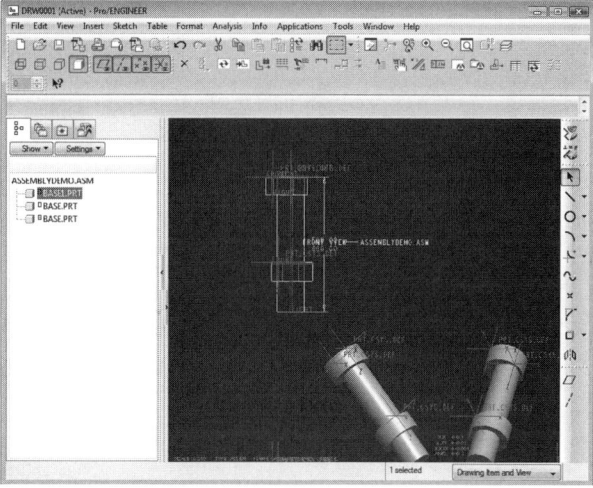

FIGURE 6.92

Now, we have completed the process to create notes in the Drawing file. With this topic, we have reached the end of this chapter.

Finally, let's have a glance at the topics we covered in this chapter.

SUMMARY

In this chapter, you learned about:

- The process to start the **Drawing** mode
- The mechanism and procedure to generate the various drawing views such as **General** view, **Projection** view, **Detailed** view, **Auxiliary** view, and **Revolved** view
- Various ways of modifying the drawing views
- Adding dimensions and tolerances to the drawing views
- The mechanism to modify an existing dimension of the drawing view
- Creating notes in drawing views to provide additional information

7

IMPLEMENTING SURFACE MODELING

In This Section

◊ Creating an Extruded Surface
◊ Creating the Sweep Surface
◊ Creating Datum Curves
◊ Creating Surfaces from Boundary Curves
◊ Merging Surfaces
◊ Trimming Surfaces

Surface modeling in Pro/ENGINEER Wildfire 4.0 is used to create and manipulate surfaces. A surface is defined as the outer body of a model having neither mass nor thickness.

In this chapter, we study how to create and manage surfaces. The chapter also discusses the process to create complex surfaces using datum curves. Sometimes, in order to get the desired surface model, we may need to merge and trim surfaces. The processes to merge and trim surfaces are also discussed in this chapter.

7.1 CREATING AN EXTRUDED SURFACE

The process to create an extruded surface is similar to the process of creating an extruded feature, which is discussed in Chapter 4, "Exploring Pro/ENGINEER Wildfire 4.0 Part Mode". The only difference between an extruded feature and an extruded surface is that an extruded feature has properties such as mass and thickness, while an extruded surface does not have these properties. Let's follow these steps to create an extruded surface:

1. *Start* the **Part** mode.

Note: The process to start the **Part** mode is discussed in Chapter 4, "Exploring Pro/ENGINEER Wildfire 4.0 Part Mode".

2. *Click* the **Extrude** (⊡) button in the **Base Features toolbar** in the **right tool chest**. The **Extrude** dashboard with the options to create an extruded surface appears, as shown in **Figure 7.1**.

3. Now, *click* the **Extrude as surface** (⊡) button in the **Extrude** dashboard to create an extruded surface (Figure 7.1).

Note: By default, the **Extrude as solid** (⊡) button is selected.

---

4. *Select* the **Placement** tab in the **Extrude** dashboard, as shown in Figure 7.1:

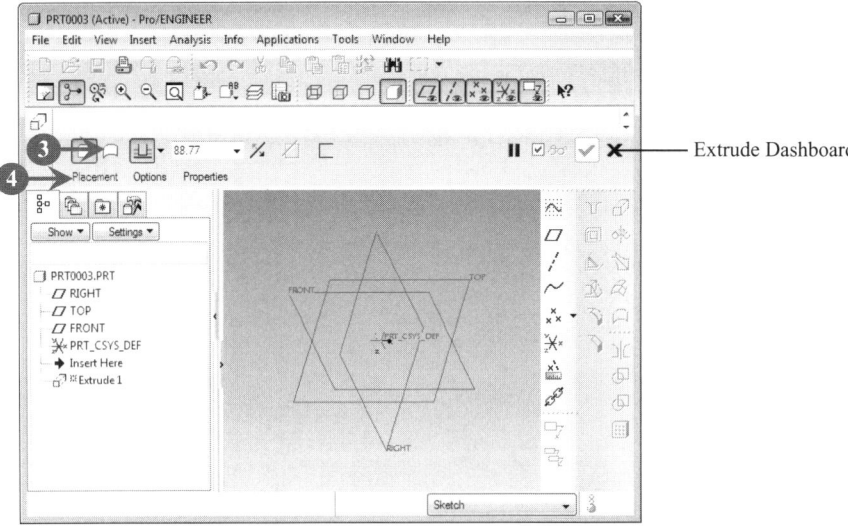

FIGURE 7.1

A slide-down panel that prompts you to select the sketching plane appears (**Figure 7.2**).

5. *Click* the **Define** button in the slide-down panel, as shown in Figure 7.2:

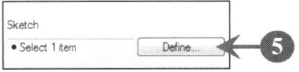

FIGURE 7.2

The **Sketch dialog box** where you have to specify the **Sketching** plane, the **Reference** plane, and the **Orientation** plane appears, as shown in **Figure 7.3**:

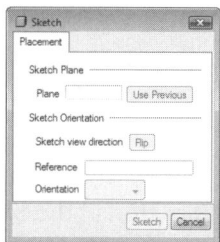

FIGURE 7.3

6. *Select* the **Front datum** plane, either in the **Model Tree** or in the **Drawing area** (Figure 7.1). The selected **Datum** plane, **Reference** plane, and **Orientation** plane automatically appears in the **Sketch dialog box**, as shown in **Figure 7.4**.

7. *Click* the **Sketch** button in the **Sketch dialog box** (Figure 7.4) to start creating the sketch of the extruded surface:

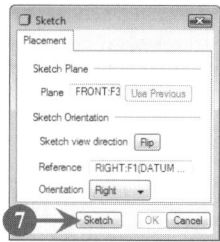

FIGURE 7.4

The **Sketcher tools toolbar** appears, as shown in **Figure 7.5**:

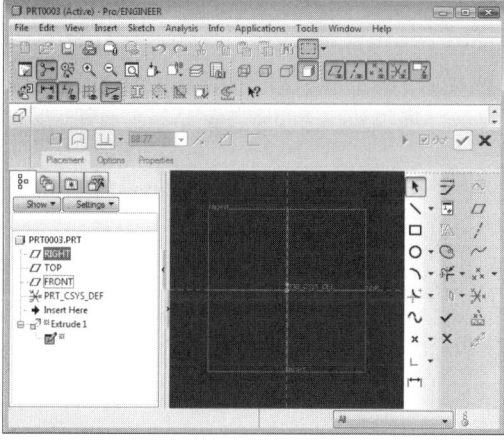

FIGURE 7.5

8. *Draw* a rectangle in the **Drawing Area**, as shown in **Figure 7.6**.

Note: The process to draw a rectangle is discussed in Chapter 3, "Exploring Pro/ENGINEER Wildfire 4.0 Sketch Mode".

9. Now, *click* the **Done** (✓) button in the **Sketcher tools toolbar** to complete the rectangle, as shown in Figure 7.6:

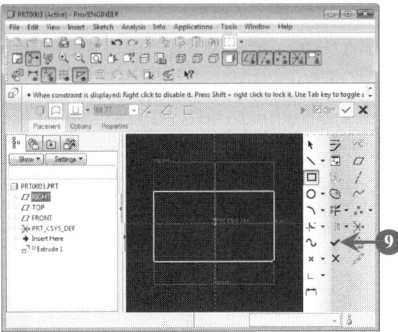

FIGURE 7.6

The rectangle appears, as shown in **Figure 7.7**.

10. *Enter* 100.00 in the **Depth box** in the **Extrude** dashboard (Figure 7.7) to specify the depth of the rectangle.

Note: You can also continue with the default depth value.

11. *Press* the **ENTER** key to project the new depth value in the **Depth box**.

12. *Click* the **Build feature** (✓) button to complete the process, as shown in Figure 7.7:

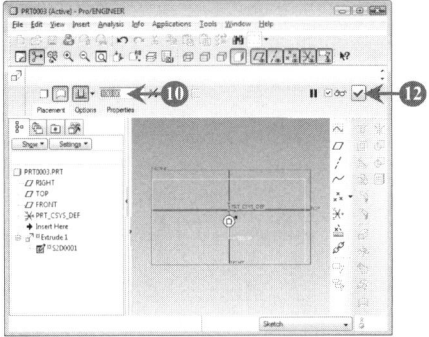

FIGURE 7.7

The extruded surface is highlighted, as shown in **Figure 7.8**:

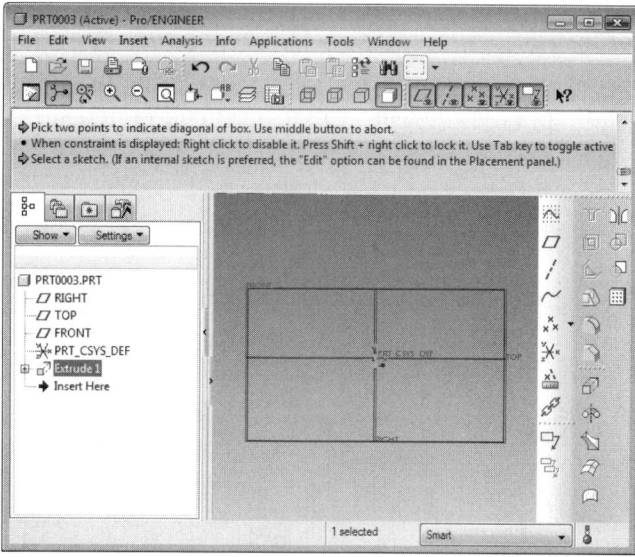

FIGURE 7.8

13. Now, press and *hold* the middle **mouse** button and spin the extruded surface to view it from different angles. The extruded surface appears, as shown in **Figure 7.9**:

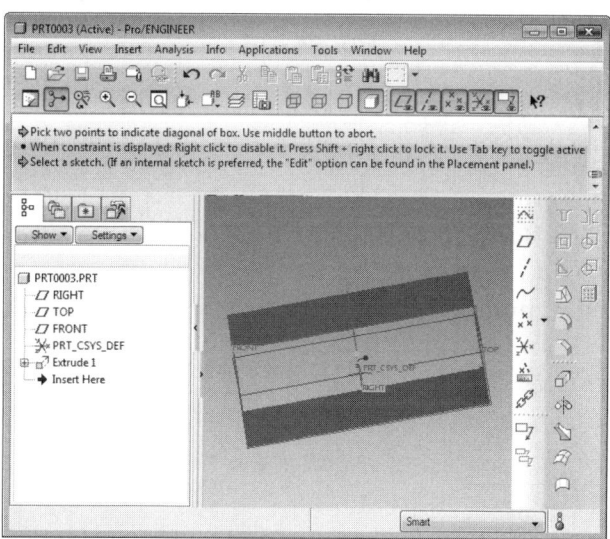

FIGURE 7.9

14. *Save* the file with the name EXTRUDEDSURFACE.prt.

In Pro/ENGINEER Wildfire 4.0, a surface can be open-ended or capped-ended. The difference between an open-ended surface and a capped-ended surface is that while an open-ended surface is open from both the sides, both the ends are capped in the capped-ended surface; in a capped-ended surface, both the ends are capped, as shown in **Figure 7.10**:

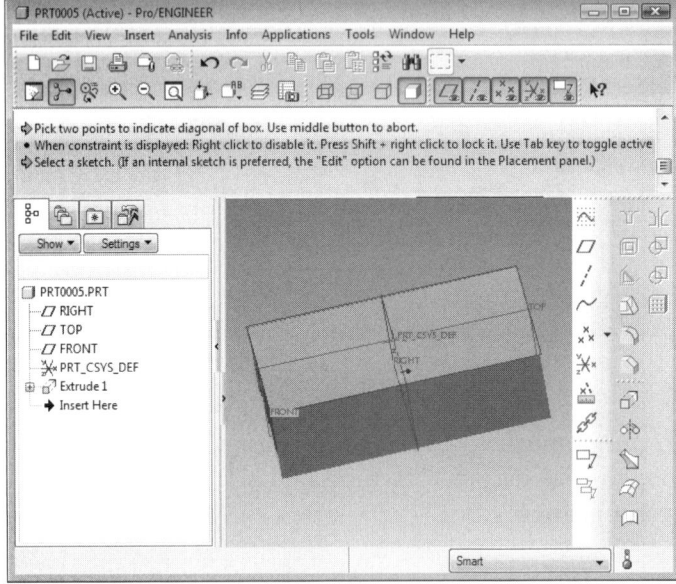

FIGURE 7.10

The process to create a capped-ended extruded surface is similar to the process of creating an open-ended extruded surface, except the following steps:

15. *Click* the **Option** tab in the **Extrude** dashboard (Figure 7.1) before clicking the **Build feature** button. A slide-down panel appears, as shown in **Figure 7.11**.

16. *Select* the **Capped ends** checkbox in the slide-down panel (Figure 7.11):

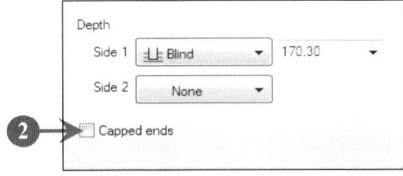

FIGURE 7.11

This enables the **Capped ends** feature in Pro/ENGINEER Wildfire 4.0 that results in the surface with capped ends (Figure 7.10). Next, let's create another surface called the sweep surface.

7.2 CREATING THE SWEEP SURFACE

A sweep surface in Pro/ENGINEER Wildfire 4.0 is created by extruding a section of an entity along a trajectory. The process to create a sweep surface is similar to the process of creating a sweep feature. The only difference between a sweep surface and a sweep feature is that a sweep surface has no mass and thickness, while a sweep feature has these properties. The process to create a sweep feature is explained in Chapter 4, "Exploring Pro/ENGINEER Wildfire 4.0 Part Mode". In this section, we study the process to create a sweep surface. Following are the steps that help you to create a sweep surface:

1. *Start* the **Part** mode.
2. Now, *select* **Insert > Sweep > Surface > Surface,** as shown in **Figure 7.12**:

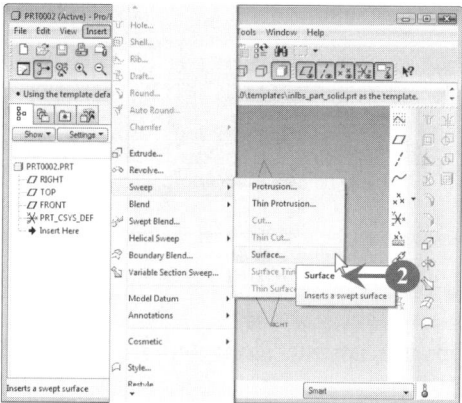

FIGURE 7.12

The **SURFACE: Sweep dialog box** (**Figure 7.13**), along with a **Menu Manager menu** (**Figure 7.14**), appears:

FIGURE 7.13

The **Menu Manager menu** provides two options:

- To sketch a trajectory
- To select a trajectory from the existing feature

In our case, we will sketch a trajectory.

3. *Select* the **Sketch Traj** option (Figure 7.14) to sketch a trajectory, as shown in Figure 7.14:

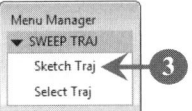

FIGURE 7.14

The appearance of the **Menu Manager menu** changes, as shown in **Figure 7.15**:

FIGURE 7.15

A **Select dialog box** also appears along with the **Menu Manager menu**, as shown in **Figure 7.16**:

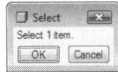

FIGURE 7.16

The **Select dialog box** prompts you to select an item, which means to select the datum plane to define a trajectory along it.

4. *Select* the **Front Datum** plane, either in the **Drawing Area** or in the **Model Tree** in the **Part** mode window. **A red arrow** appears in the direction of viewing the selected datum plane, as shown in **Figure 7.17**:

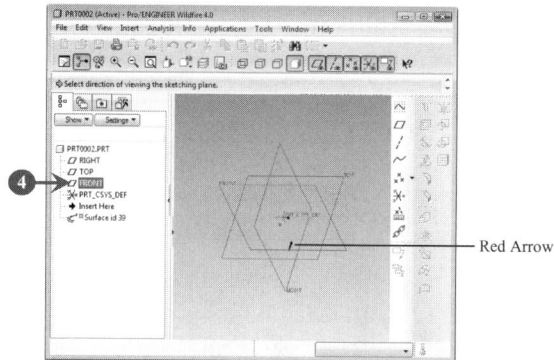

Red Arrow

FIGURE 7.17

Appearance of the **Menu Manager menu** changes once again, as shown in **Figure 7.18**.

5. *Select* the **Okay** option to continue creating a sweep feature, as shown in **Figure 7.18**:

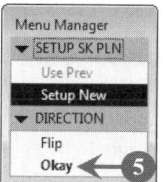

FIGURE 7.18

The appearance of the **Menu Manager menu** changes again, as shown in **Figure 7.19**.

Now, you have to select the orientation view of the sweep feature from the options given in the **SKET VIEW** submenu of the **Menu Manager menu** (Figure 7.19).

6. *Select* the **Top** option in the **SKET VIEW** submenu, as shown in Figure 7.19:

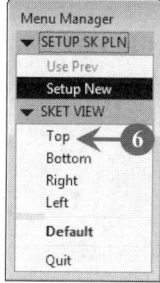

FIGURE 7.19

The appearance of the **Menu Manager menu** changes again, as shown in **Figure 7.20**:

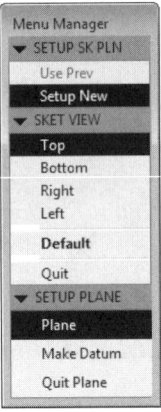

FIGURE 7.20

The **Select dialog box** (Figure 7.16) appears again. Both the **Menu Manager menu** (Figure 7.20) and the **Select dialog box** prompt you to select the **Reference** plane of the **Front Datum** plane in the **Part** mode window.

7. *Select* the **Top Datum** plane in the **Model Tree**. The **Sketcher tools toolbar** appears to let you sketch a trajectory (**Figure 7.21**).

8. *Select* the **3-Point/Tangent End** (⟍▾) button in the **Sketcher tools toolbar** to draw an arc, as shown in Figure 7.21:

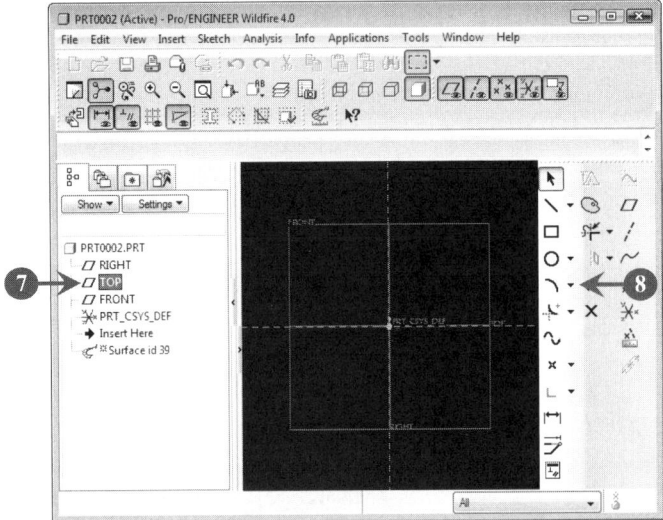

FIGURE 7.21

A **yellow arrow** is attached to the arc. This yellow arrow represents the start point of the trajectory and points in the direction in which the sweep surface will be created.

Note: The process to draw an arc is explained in Chapter 3, "Exploring Pro/ENGINEER Wildfire 4.0 Sketch Mode".

9. *Click* the **Done** (✔) button to complete the sketch of the trajectory, as shown in **Figure 7.22**:

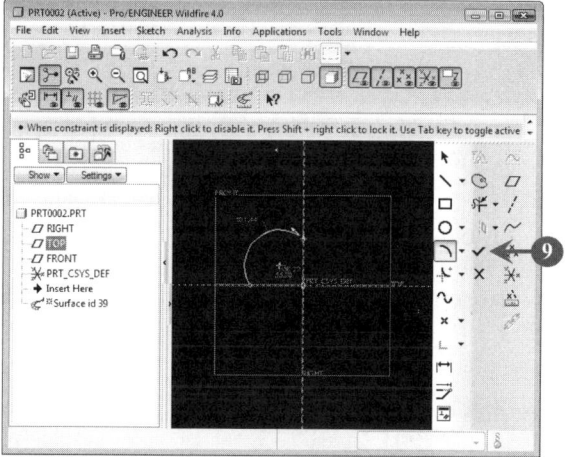

FIGURE 7.22

The view of the **Drawing Area** is changed to let you create a section for the sweep surface, as shown in **Figure 7.23**:

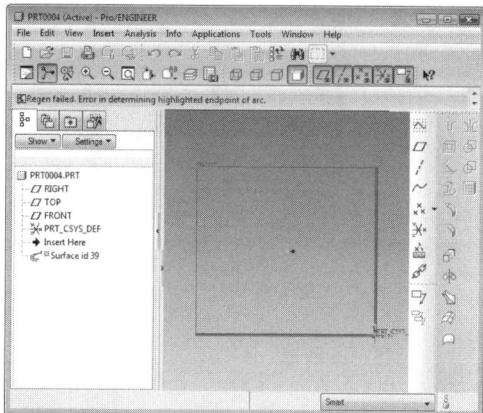

FIGURE 7.23

The appearance of the **Menu Manager menu** (Figure 7.20) again changes, as shown in **Figure 7.24**.

Now, the **Menu Manager menu** (Figure 7.24) lets you select the option to create either an open-ended sweep feature or a capped-ended sweep feature. Select the **Open Ends** option to create an open-ended sweep feature, or the **Capped Ends** option for a capped-ended feature. By default, the **Open Ends** option is selected. We continue with the default option.

10. *Select* the **Done** option in the **Menu Manager menu** (Figure 7.24) to create an open-ended sweep feature:

FIGURE 7.24

The view of the **Drawing Area** changes once again (**Figure 7.25**).

11. *Click* the **Center and Point** (○ ⁻) button in the **Sketcher tools toolbar** (Figure 7.25):

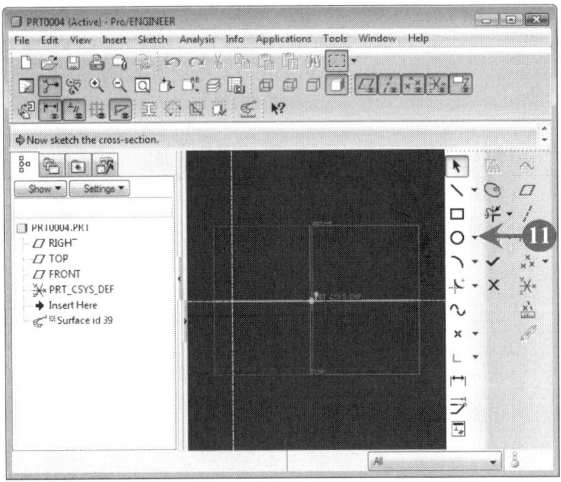

FIGURE 7.25

In Figure 7.25, the intersection point of two infinite yellow lines represents the starting point of the trajectory. The start point of the sweep section should lie at the start point of the trajectory.

A circle is drawn as a sweep section, as shown in **Figure 7.26**.

Note: The process to draw a circle is explained in Chapter 3, "Exploring Pro/ENGINEER Wildfire 4.0 Sketcher Mode".

12. *Click* the **Done** (✔) button in the **Sketcher tools toolbar** (Figure 7.26):

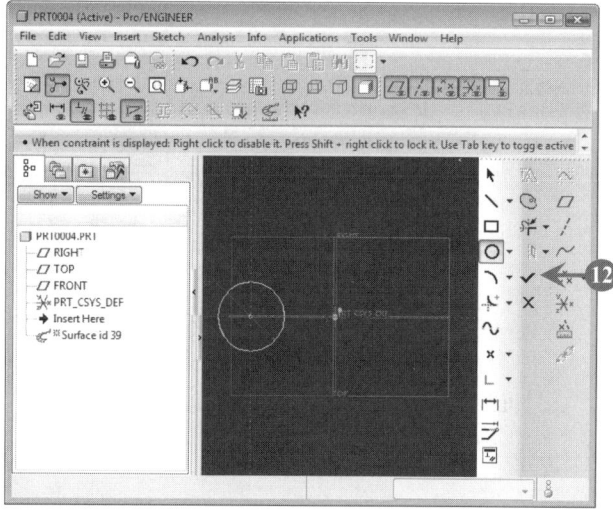

FIGURE 7.26

The sweep surface is created, as shown in **Figure 7.27**:

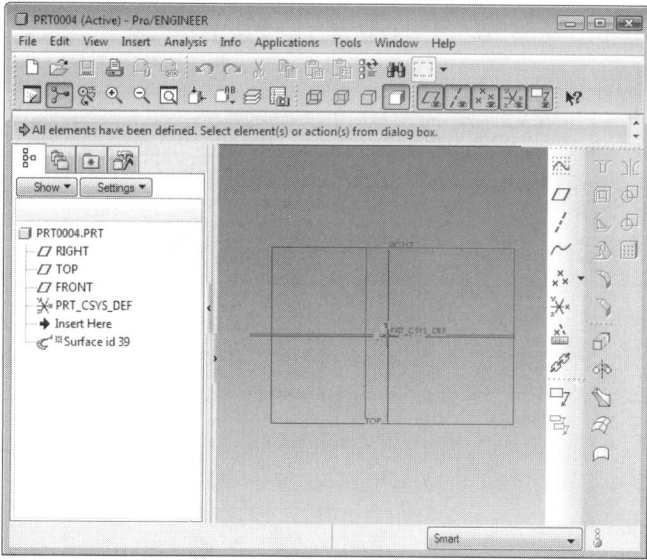

FIGURE 7.27

13. *Click* the **Preview** button in the **SURFACE: Sweep dialog box** (Figure 7.13). The sweep surface appears, as shown in **Figure 7.28**.

14. Now, *select* the sweep surface, as shown in Figure 7.28:

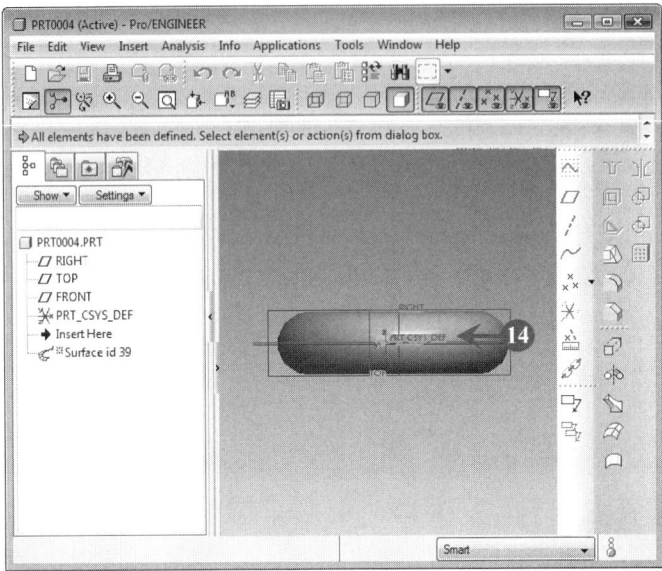

FIGURE 7.28

15. *Press* and *hold* the middle **mouse** button and *spin* the sweep surface to view it from different angles, as shown in **Figure 7.29**:

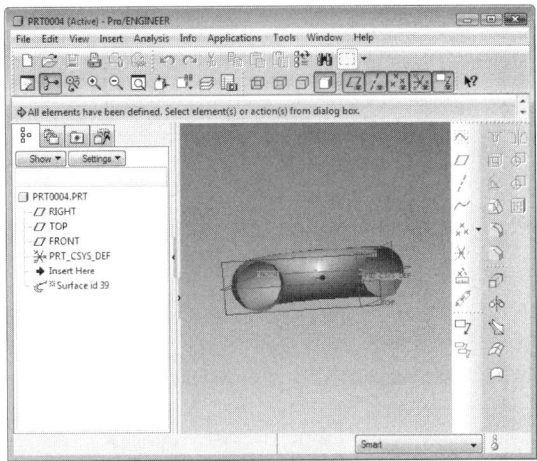

FIGURE 7.29

16. *Click* the **OK** button in the **SURFACE: Sweep dialog box** (Figure 7.13) to finish the process. The sweep surface is highlighted in red, as shown in **Figure 7.30**:

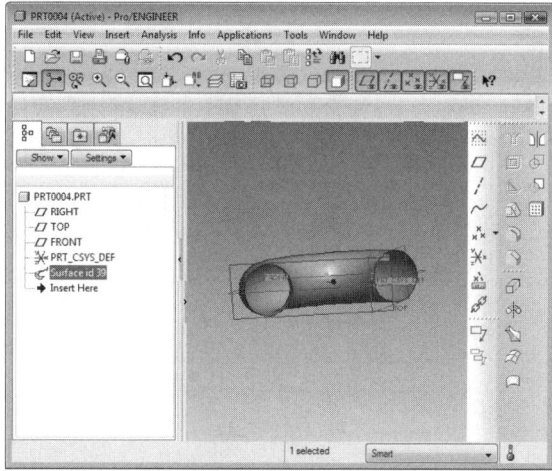

FIGURE 7.30

17. Now, *save* the file. Here, we saved the file with the name SWEEPSURFACE.prt.

Till now, we have studied how to create a sweep surface. Next, let's discuss how to create datum curves.

7.3 CREATING DATUM CURVES

Datum curves are defined as the curves having no mass, volume, or thickness. Datum curves are used to create complex surfaces that are not easy to create with the datum planes. Once a datum curve is created, it appears in the **Model Tree**. A datum curve in Pro/ENGINEER Wildfire 4.0 is created using the **Curve** ($\sim$) button in the **Datum toolbar**. A datum curve is created along with a surface; therefore, we use the extruded surface created in the "Creating an Extruded Surface" section of this chapter to create a datum curve. Let's follow these steps to create a datum curve:

1. *Start* the **Part** mode.
2. *Open* the EXTRUDEDSURFACE.prt file. The EXTRUDEDSURFACE.prt file opens (**Figure 7.31**).

3. Now, *click* the **Curve** (~) button in the **Datum toolbar**, as shown in Figure 7.31:

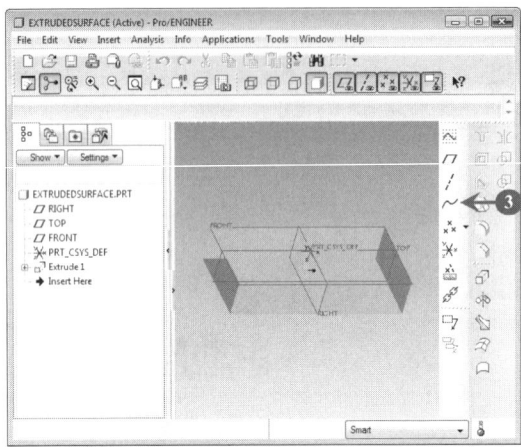

FIGURE 7.31

Note: The process to open a .prt file is explained in Chapter 4, "Exploring Pro/ENGINEER Wildfire 4.0 Part Mode".

The **Menu Manager menu** appears, as shown in **Figure 7.32**. It contains the following options:

- **Thru Points:** Creates a datum curve along the specified points on a surface.
- **From File:** Creates a datum curve by importing the geometry of a model. The imported geometry then automatically gets converted into a datum curve.
- **Use Xsec:** Creates a datum curve that has geometry of an existing cross section.
- **From Equation:** Creates a datum curve through specifying an equation using the coordinate systems.

By default, the **Thru Points** option is selected. In this section, we continue with the default option.

4. *Select* the **Done** option in the **Menu Manager menu** (Figure 7.32) to continue creating a datum curve:

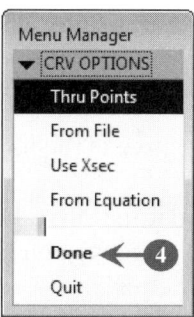

FIGURE 7.32

The **CURVE: Through Points dialog box** appears, as shown in **Figure 7.33**:

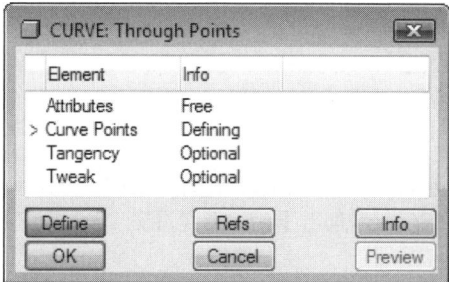

FIGURE 7.33

This dialog box allows you to select the points on a surface. Apart from this, the appearance of the **Menu Manager menu** changes, as shown in **Figure 7.34**:

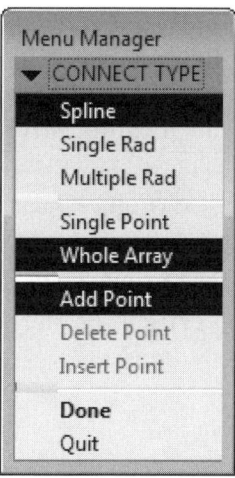

FIGURE 7.34

The changed **Menu Manager menu** (Figure 7.34) contains the following options according to which the shape of the curve is defined:

- **Spline:** Creates a datum curve in the form of a spline. This option is selected by default.
- **Single Rad:** Creates a datum curve having constant radius between two points. If there are more than two points, then the datum curve between all the points has the same radius.
- **Multiple Rad:** Creates a datum curve by specifying different radii between the points.
- **Single Point:** Selects a single point on a surface individually. By selecting the points individually, multiple datum curves can be created between the selected points.
- **Whole Array:** Selects all the points to create a single datum curve. This option is selected by default.

In this section, we continue with the default options.

Apart from the **CURVE: Through Points dialog box**, the **Select dialog box** also appears, as shown in **Figure 7.35**:

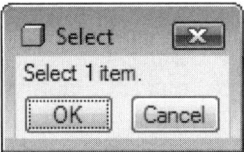

FIGURE 7.35

The **Select dialog box** prompts you to select an item, which means to select points on the surface to create a datum curve.

5. *Select* the extruded surface (Figure 7.31).
6. *Press* and *hold* the middle **mouse** button and *spin* the extruded surface in the way that it appears in **Figure 7.36**.
7. Now, *select* all four vertices of the extruded surface, as shown in Figure 7.36:

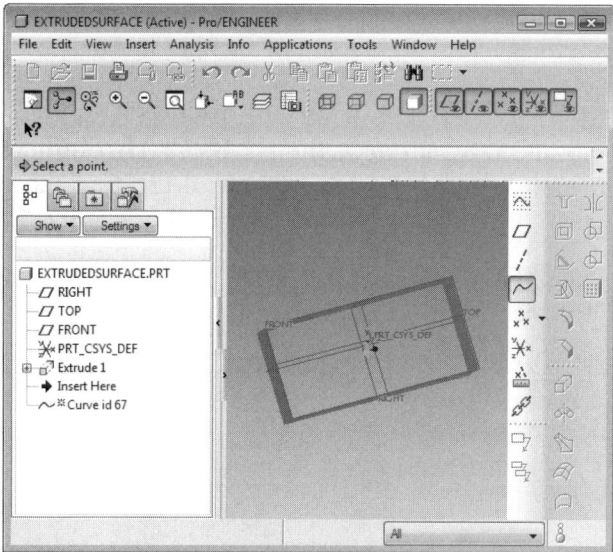

FIGURE 7.36

A datum curve is outlined, as shown in **Figure 7.37**:

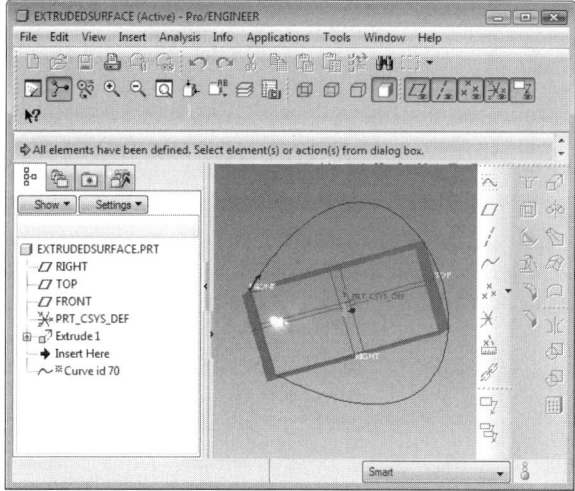

FIGURE 7.37

In Figure 7.37, the datum curve is outlined in blue. A blue arrow attached to the datum curve indicates the start point of the datum curve.

8.   *Select* the **Done** option in the **Menu Manager menu** (Figure 7.34).

9.   Now, *click* the **OK** button in the **CURVE: Through Points dialog box** (Figure 7.33) to complete the process of creating the datum curve. The datum curve is highlighted in red, as shown in **Figure 7.38**:

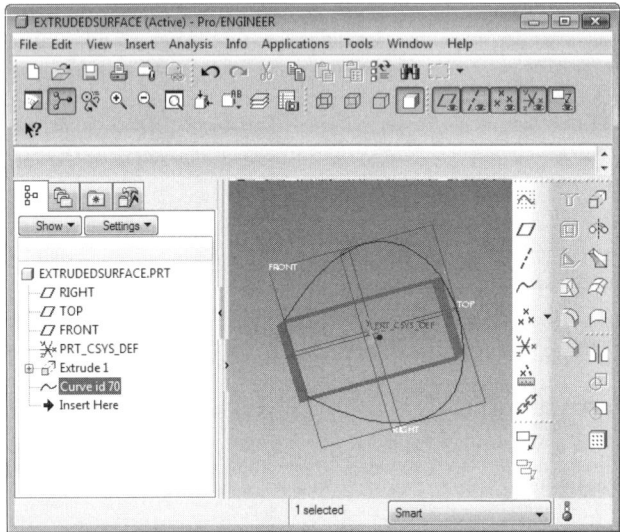

FIGURE 7.38

10.   *Save* the file with the name CURVE.prt.

Note: You can also create a datum curve along with a feature to create a complex feature.

With this, we have completed the process of creating a datum curve. Next, let's see how to create a surface by blending the datum curve with an existing surface.

7.4 CREATING SURFACES FROM BOUNDARY CURVES

In the previous section, we created a datum curve along the extruded surface. In this section, we study the process of blending the datum curve with a surface. To blend a datum curve with a surface to create another surface, the **Boundary Blend** dashboard is used. Following are the steps that help you to create a surface by blending a datum curve with another surface:

1. *Start* the **Part** mode.
2. *Open* the CURVE.prt file. The CURVE.prt file opens (**Figure 7.39**).

Note: If the orientation of the file is not similar to the orientation, as shown in Figure 7.39, then select the **BACK** orientation available in the **Named View List** option in the **View toolbar.**

3. Now, *click* the **Boundary Blend** ( ) button in the **Base features toolbar**, as shown in Figure 7.39:

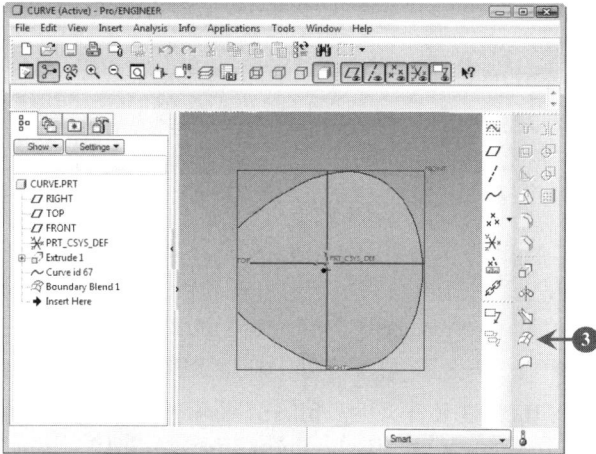

FIGURE 7.39

The **Boundary Blend** dashboard appears, as shown in **Figure 7.40**:

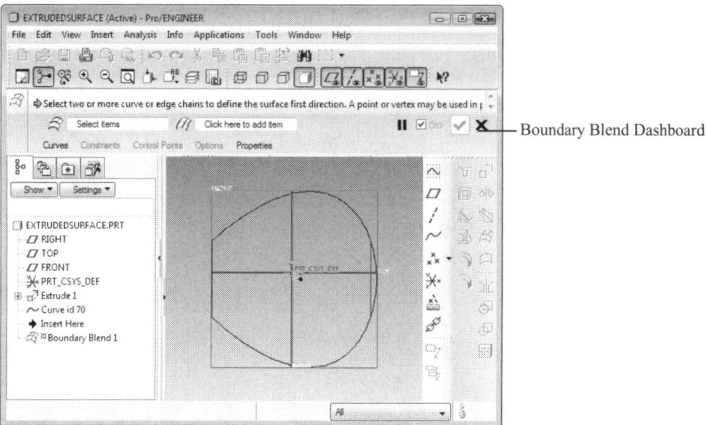

FIGURE 7.40

Now, you need to select the curve and the edges of the extruded surface to blend them.

4. *Select* an edge of the extruded surface, as shown in **Figure 7.41**:

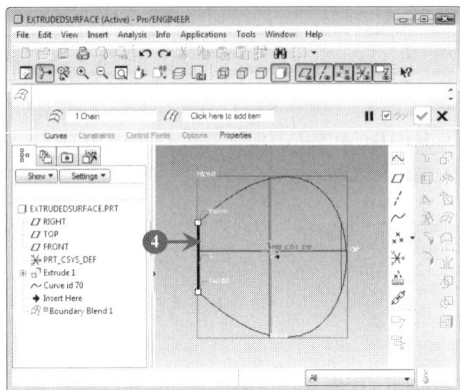

FIGURE 7.41

The selected edge is highlighted in red. The selected edge also appears in the **Boundary Blend** dashboard in the left edit box. In the **Boundary Blend** dashboard, the selected edge is numbered as **1 Chain**. Now, if you select more than one edge, then it will be successively numbered.

5. Now, *select* the datum curve while pressing the **CTRL** key. The surface that is created after blending is shown in **Figure 7.42**.

As soon as the curve is selected, it is highlighted in red (Figure 7.42). The total number of selected items now becomes two **Chains** in the **Boundary Blend** dashboard.

6. Now, *click* the **Build feature** (✔) button to complete the process, as shown in Figure 7.42:

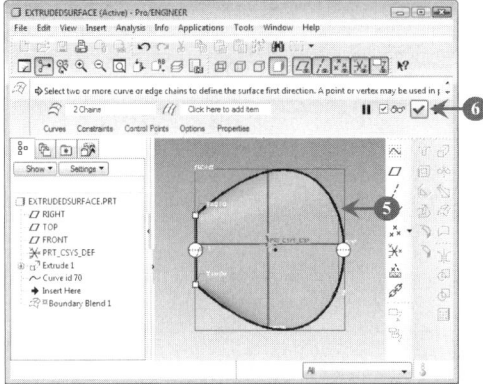

FIGURE 7.42

The new surface appears, as shown in **Figure 7.43**:

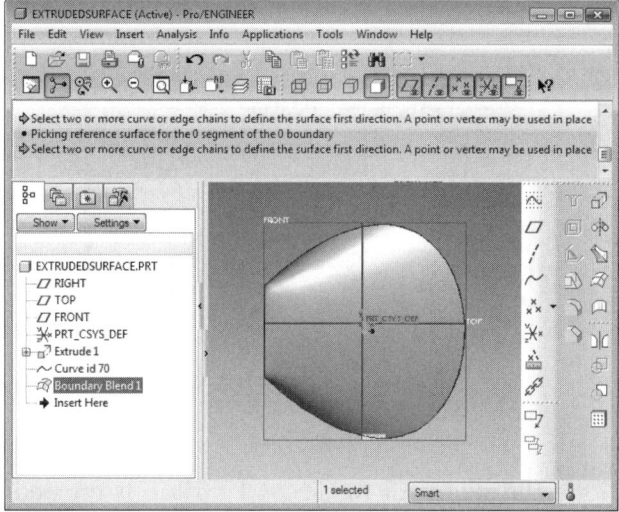

FIGURE 7.43

Note: Use the options available in the **Named View List** option in the **View toolbar** to view the surface in different orientations.

---

7. *Save* the file with the name BLENDEDCURVE.prt.

Till now, we have studied the process to create a surface by blending curves. Next, let's study how to merge surfaces.

7.5 MERGING SURFACES

In Pro/ENGINEER Wildfire 4.0, merging is defined as the process to join or intersect two surfaces to create a single surface. The surfaces that have to be merged should be attached at their edges. In Pro/ENGINEER Wildfire 4.0, the **Merge** dashboard is used to merge two surfaces. The **Merge** dashboard is invoked by clicking the **Merge** (⌷) button in the **Edit Features toolbar**. Along with merging, the surfaces are also automatically trimmed. Let's follow these steps to merge two surfaces:

1. *Start* the **Part** mode.
2. *Open* the EXTRUDEDSURFACE.prt file (Figure 7.9). The EXTRUDED-SURFACE.prt file works as the first surface.

3. Now, *create* the second surface along with the extruded surface, as shown in **Figure 7.44**:

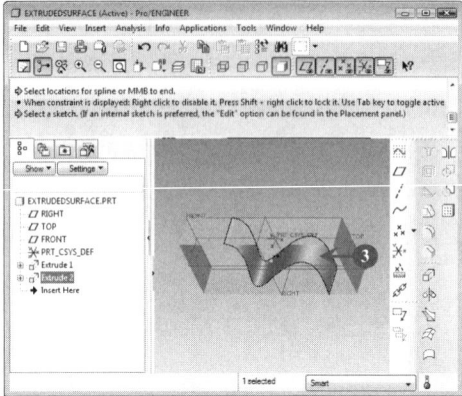

FIGURE 7.44

In Figure 7.44, the sketch of the second surface is created using the **Spline** button in the **Sketcher tools toolbar**.

Note: The process to draw a sketch using the **Spline** button is discussed in Chapter 3, "Exploring Pro/ENGINEER Wildfire 4.0 Sketch Mode".

4. Now, *hold* the **CTRL** key and *select* both surfaces that are to be merged. Selected surfaces are highlighted in red. The **Merge** button also becomes enabled.

Note: By default, the **Merge** button is disabled. The **Merge** button is enabled only when the surfaces to be merged are selected.

5. *Click* the **Merge** button, as shown in **Figure 7.45**:

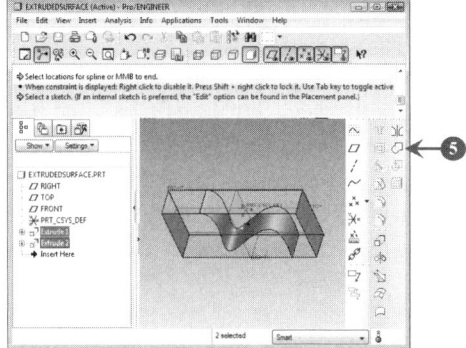

FIGURE 7.45

The **Merge** dashboard appears (**Figure 7.46**).

The yellow dots in Figure 7.46 represent the area that is retained after merging the surfaces. The area that is not covered with the yellow dots is trimmed.

The yellow arrows are pointing in the direction in which the surfaces are merged. You can also change the direction of these arrows by clicking the **Change side of first quilt to keep** (✗) and **Change side of second quilt to keep** (✗) buttons in the **Merge** dashboard.

6. Now, *click* the **Build feature** (✔) button to complete the process, as shown in Figure 7.46:

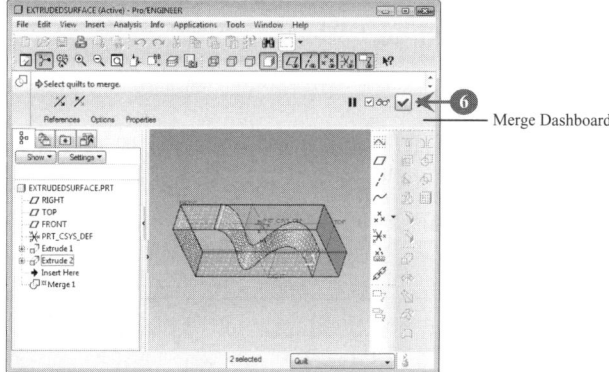

Merge Dashboard

FIGURE 7.46

Note: In Pro/ENGINEER Wildfire 4.0, you can merge surfaces either by intersecting them or by joining them. Both the **Intersect** and **Join** options appear when the **Options** tab in the **Merge** dashboard is selected. By default, the **Intersect** option is selected.

In our case, we continue with the default option. The output is shown in **Figure 7.47**:

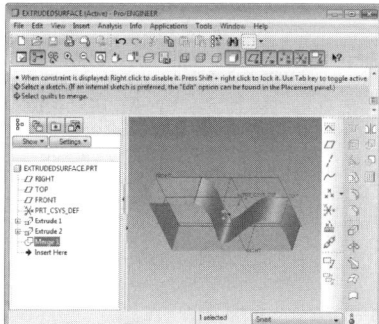

FIGURE 7.47

7. *Save* the file with the name MERGEDSURFACE.prt.

After discussing the merging process, let's next study how to trim surfaces.

7.6 TRIMMING SURFACES

Trimming in Pro/ENGINEER Wildfire 4.0 is used to crop the surfaces. Trimming is done by using the **Trim** dashboard. However, the **Trim** dashboard is invoked by clicking the **Trim** ( ) button in the **Edit Feature toolbar**. Here, we use the sweep surface created in the section, "Creating a Sweep Surface" in this chapter to trim. Let's now follow these steps to understand the process of trimming surfaces:

1. *Start* the **Part** mode.
2. *Open* the SWEEPSURFACE.prt (Figure 7.30) file to access the sweep surface created in the section "Creating a Sweep Surface".
3. Now, *select* the sweep surface that you want to trim (**Figure 7.48**).

Note: By default, the **Trim** button is disabled. The **Trim** button is enabled only when the surface to be trimmed is selected.

---

4. *Click* the **Trim** ( ) button in the **Edit Feature toolbar**, as shown in Figure 7.48:

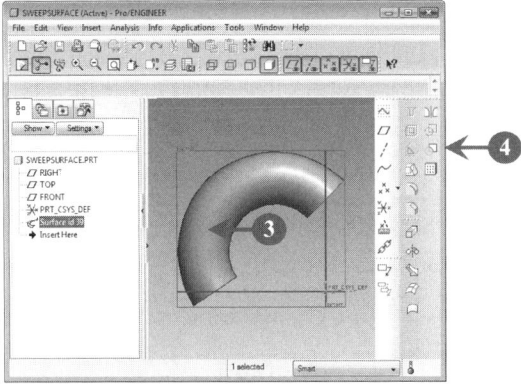

FIGURE 7.48

The **Trim** dashboard appears, as shown in **Figure 7.49**:

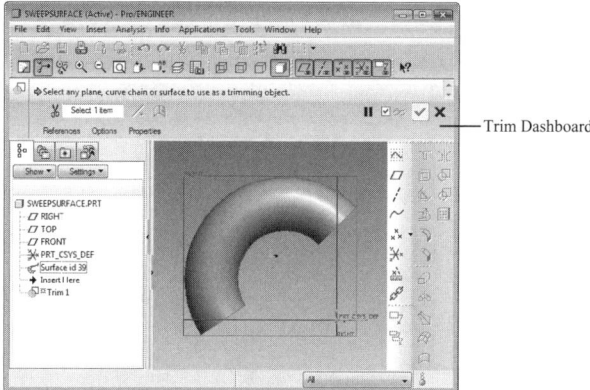

FIGURE 7.49

In Figure 7.49, you can see a statement written in the message area. This statement prompts you to select datum plane, curve chain, or surface to use as a trimming object.

5. Now, *select* the **Top Datum** plane in the **Model Tree** (**Figure 7.50**).

The selected portion gets highlighted with yellow dots. These dots represent the area that will be retained after trimming. Besides the highlighted area, there is a yellow arrow that represents the direction in which the surface is trimmed. You can change the direction by clicking the **Flip** (✗) button in the **Trim** dashboard.

6. *Click* the **Build feature** (✓) button to complete the process, as shown in Figure 7.50:

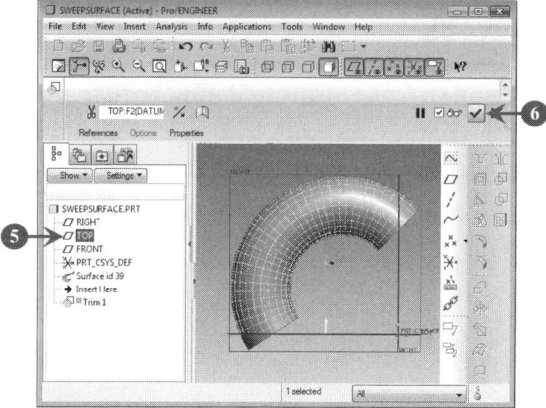

FIGURE 7.50

The trimmed surface appears, as shown in **Figure 7.51**:

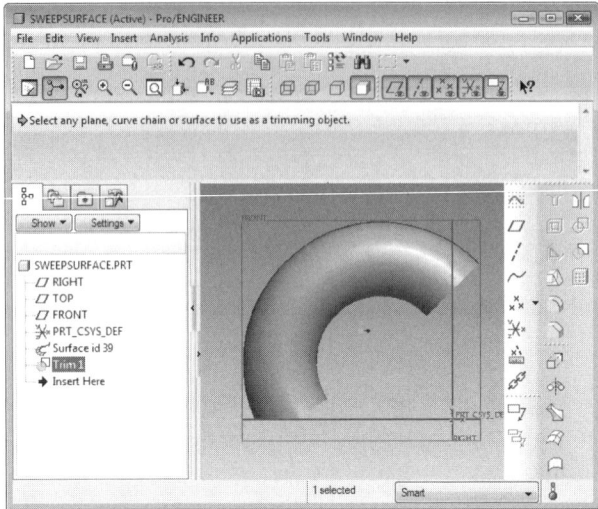

FIGURE 7.51

7. *Save* the file with the name TRIMMEDSURFACE.prt.

In this way, surfaces are trimmed in Pro/ENGINEER Wildfire 4.0. Let's now summarize the topics covered in this chapter.

SUMMARY

In this chapter, you have learned about:

- Creating an extruded surface
- Creating a sweep surface
- Designing complex surfaces using datum curves
- Blending datum curves and surfaces
- Merging two surfaces to create a single complex surface
- Trimming a surface

INDEX